BOOM!

BOOM!

THE CHEMISTRY AND HISTORY OF EXPLOSIVES

SIMON QUELLEN FIELD

CHICAGO
REVIEW
PRESS

Published by Chicago Review Press Incorporated
814 North Franklin Street
Chicago, Illinois 60610
ISBN 978-1-61373-805-4

Library of Congress Cataloging-in-Publication Data
Names: Field, Simon (Simon Quellen)
Title: Boom! : the chemistry and history of explosives / Simon Quellen Field.
Description: Chicago, Illinois : Chicago Review Press, [2017] | Includes index.
Identifiers: LCCN 2016047898 (print) | LCCN 2016049739 (ebook) | ISBN 9781613738054 (pbk. : alk. paper) | ISBN 9781613738061 (pdf) | ISBN 9781613738085 (epub) | ISBN 9781613738078 (kindle)
Subjects: LCSH: Explosives—Popular works. | Explosives—History—Popular works. | Chemical weapons—History—Popular works.
Classification: LCC TP270.5 .F54 2017 (print) | LCC TP270.5 (ebook) | DDC 662/.2—dc23
LC record available at https://lccn.loc.gov/2016047898

Cover design: John Yates at Stealworks
Cover image: Getty Images/Dimitri Otis
Interior design: Jonathan Hahn

Printed in the United States of America
5 4 3 2 1

To Connor

CONTENTS

INTRODUCTION

Bang!

I always like to start a book off with a bang.

This is a book about bangs. Big ones but also little ones. The tiny bang of a kernel of popcorn exploding into a white fluffy treat. The thousands of little bangs every second that power internal combustion engines. The little bangs of firecrackers and cap guns. The big bangs of lightning and thunder. The explosions of mining, warfare, volcanoes, and collapsing stars.

While many of those explosions are natural, this book is mainly about human-made explosions—the discoveries that enable them, the people who made the discoveries, and the consequences for good and ill. Not just the who, when, and why but also the how. How did Chinese alchemists come to create black powder? What accidents led to the discovery of high explosives? How do explosives actually work?

There are three groups of explosions: mechanical explosions, chemical explosions, and nuclear explosions.

Mechanical explosions happen when pressure builds up in an enclosed space until the container suddenly bursts to release the pressure, like that kernel of popcorn. It is heated until the water inside turns to steam. The pressure of the steam continues to increase as it is heated until the tough little shell around it bursts. Lightning is a mechanical explosion. The electric current in the lightning bolt heats the air to over 50,000°F in less than ten microseconds. The resulting hot air expands at a rate faster than the speed of sound. In this case, the "enclosed space" is caused by inertial confinement, which just means the heating happens so fast, the expanding gas can't get out of the way fast enough.

Chemical explosions involve fire. They can be reactions that are so fast they mimic lightning in their ability to heat gases faster than they can escape, or they can be slower (but still generally fast) reactions that build up pressure in a container.

1

Black powder in a musket or firecracker is one example of a reaction that explodes if contained but merely burns in the open. Another is the explosions that happen in an internal combustion engine. It is difficult to make fuels in open air burn fast enough to explode. This is why fuel-air bombs (also called thermobaric weapons) were not invented until World War II and not used much until the Vietnam War in the 1960s.

Accidental chemical explosions happen when fuel and air mix in the right proportions. Coal dust explosions in the confines of a coal mine or flour dust explosions in mills are both deadly examples. Dust explosions are less common without an enclosing mine or building.

The key to chemical explosions is to mix the fuel and the oxidizer (such as air) very well, so that many fuel molecules are in contact with their respective oxidizer molecules. A pile of coal dust will not explode. The fuel and the air do not mix well enough. If the dust is not fine enough, an explosion is less likely. Again, too much of the fuel surrounds more fuel and is not close to the oxidizer.

A second key is the right proportion of fuel to oxidizer. If the fuel is surrounded by too much air, or vice versa, only a part of the mix will burn. The parts that remain merely soak up heat that could have gone into igniting more reactants and absorb some of the kinetic energy that could have gone into bursting the container wall. Ideally, each atom of fuel pairs with an atom of oxidizer, so the reaction goes to completion.

This second concept leads to the definition of lower and upper explosive limits. If there is not enough fuel, known as a "lean" mixture, a gas or vapor will not burn. If there is too much fuel, a "rich" mixture, again it will not burn. Some fuels have very narrow limits. Gasoline, for example, will only burn between about 1% to 7% fuel in air. Methane (natural gas), propane, and butane also have narrow combustion ranges. This is why they are safer fuels to use than, say hydrogen, which has a range of 4% to 75%, meaning that almost any amount of hydrogen in air will at least burn, and quite a large range will explode.

This recipe of the right proportions of oxidizer to fuel, mixed as closely as possible, is the hallmark of chemical explosives. The history of gunpowder is all about finding the right proportions of sulfur, charcoal, and nitrates and learning how to mix them so that the particles of each are tiny and homogenous. Later, mixing them so well that the oxidizer and the fuel are on the same molecule produced explosives like guncotton and nitroglycerin.

As humans learned to make many different types of explosives, they found a need to compare them to one another. Most people have a vague understanding that dynamite is more powerful than gunpowder, but what does it mean to be more powerful? Power is energy divided by time. If more energy can be produced in the same time, there is more power. Likewise, if the same energy can be delivered in less time, there is more power. In explosives, this relates to the speed of the chemical reaction that generates the energy.

In history's first explosive, black powder, the chemical reaction happens at a speed slower than the speed of sound in the mixture. Black powder does not detonate. It *deflagrates*, which just means it burns rapidly. It still delivers quite a bit of energy, but it takes longer to do so. By containing the energy and releasing it suddenly when the container bursts, its power can increase quite a bit. But the total energy has not changed.

In many other explosives, the chemical reaction happens at a rate faster than the speed of sound in the material. A supersonic shockwave propagates through the explosive, releasing the energy faster than heat can be conducted. This is *detonation*, and the velocity of detonation relates to the power of the explosive.

Around the time of World War I, the explosive *trinitrotoluene* (better known as TNT) was in widespread use. As more explosives were developed, people needed a way to determine how much of a new explosive they would need to have the same effect they were used to getting with TNT. If an application used a pound of TNT to move rock in a mine or launch a projectile, how much of the new explosive would they need to use?

Comparing an explosive to TNT reveals what is called the *relative effectiveness* of the explosive. TNT has a velocity of detonation of 6,900 meters per second (15,435 miles per hour). It has a density of 1.6 times that of water. The dynamite Alfred Nobel invented—a mixture of 75% nitroglycerin and 25% inert material—has a detonation velocity of 7,200 meters per second (16,106 miles per hour), which is a bit over 4% faster than TNT. It has a density 1.48 times that of water, so it is less dense than TNT. These factors lead to a relative effectiveness of 1.25. To get the same effect as a pound of TNT, you need only 1/1.25 (80%) as much dynamite by weight, or 12.8 ounces.

This concept goes further when discussing nuclear weapons. Instead of relative effectiveness, terms like *kilotons* and *megatons* are used. These relate the effectiveness to thousands of tons of TNT, or millions of tons.

An explosion releases energy in the form of heat and the motion of the expanding gases. What is important is the rate at which the energy is released—the power. Energy is measured in many different ways. The energy in a pound of TNT is 1.89 megajoules, or 453 kilocalories, or 527 watt-hours, or 1.4 million foot-pounds. That's the same energy in a large candy bar. It is also about the same amount of energy you need to accelerate a car to 60 miles per hour. The difference is that the TNT releases all that energy in a microsecond or two, instead of the several seconds it takes to get the car onto the highway, or the hour it takes to run off the calories in that candy bar.

Power is energy per unit time, so it comes as no surprise that power is measured in as many ways as energy. Horsepower, watts, and foot-pounds per second are some of them. A pound of TNT would make a ball a bit over 3 inches in diameter. TNT's detonation velocity of 6,900 meters per second means that the whole pound would explode in less than six microseconds; 1.89 megajoules of energy in six microseconds is 315 billion watts. By comparison, the largest US nuclear power plant produces 3.3 billion watts and serves four million people. Of course, it does it day in and day out, not just for six microseconds.

This is a book about the *history* of explosives. Not a lot of written information from the very earliest days has survived, and much of that is from people writing about things that happened years or even centuries before the writer was born. Teasing out what is real from what is conjecture is not always easy, and I try to avoid conjecture in favor of making it clear what is unknown.

Most of the history of explosives, however, coincides with the history of chemistry, and there are excellent records of the original scientific papers on these subjects. Despite that, I have found in researching this book that there are numerous errors in what are normally thought of as trustworthy sources. I have found encyclopedia entries that state something in one paragraph and contradict it in the next. This is not limited to Internet encyclopedias, although there is much misinformation there. It is said that a man with one watch knows what time it is. Reading two biographies of a notable scientist that disagree on dates or other particulars make a writer question everything.

Where I can, I try to find the original paper by the chemist who made the discovery. These papers have dates and are often readily available—for example, the UK's Royal Society, where many of the early scientists published their work, has digitized its journal, going back centuries, and made it available for free on the Internet. In other cases, there are patents establishing a clear date for both the original filing and the granting (often years later). This does not tell how many years of work led up to the patent, but it does often establish priority and indicates when other researchers might have known about a breakthrough or a process. To be sure of accuracy, I will generally cite a patent date or the date of a paper's publication, rather than a secondhand account of when a discovery was made, no matter how good the credentials of the writer appear to be.

This is not a book full of footnotes, intended for scholarly research. However, given the name on a patent and the date it was filed, readers can quickly find the quoted material online. The same goes for papers mentioned with dates. For other information, the casual reader can use the names and molecules mentioned to search for books that mention them. Many of those books have been digitized, and in many cases, large portions of the book can be read online before purchase or a visit to the library.

BLACK POWDER

Black powder, a mixture of nitrates with sulfur and charcoal, was originally developed during the Tang dynasty in China, in the seventh century. This was the golden age in China, known for its poetry and painting, along with gunpowder and other inventions, such as wood block printing. With a population of fifty million people and vastly expanded borders, there were many resources available for the expansion of knowledge and technology.

In searching for medicines, Chinese practitioners experimented with many extracts and mixtures. They were already familiar with sulfur and charcoal, both of which have medicinal uses. They were also familiar with saltpeter, which in its impure form is a mixture of several nitrates, including sodium and potassium nitrates. Purified potassium nitrate, where enough of the sodium had been removed that the resulting crystals imparted a purple color to a flame, had been available for at least three centuries prior to its use in gunpowder.

Heating and burning mixtures was a common practice in making medicines, and heating charcoal or sulfur with nitrates generates a notable reaction. The Chinese named this mixture *huo yao*, or "fire chemical." The first known recipes for black powder, written down as early as 1044, used less nitrate than is commonly used today. The nitrate is the oxidizer, and if the mixture was to be burned in the open air, it may not have been as necessary to have a lot of oxidizer. Later uses, in which the powder was confined in rockets or guns, would lead to mixtures with about 75% nitrate.

Of the three ingredients, charcoal was likely the easiest to procure. Charcoal was used in metal smelting and other industries because it burns very hot, and the temperature can be controlled easily by controlling the flow of air. Commercial production of charcoal was an involved affair. A stack of wooden logs was carefully made and covered in sod, clay, or earth to prevent outside air from getting in, with a chimney for controlled airflow. The wood was set on fire, and once it was well started, the openings for air were closed off with more sod, clay, or earth. The whole apparatus was carefully tended for as long as five days or more, and any leaks that developed were covered to prevent the product from going up in smoke.

As a source of combustible carbon, charcoal has some special properties that make it work better in black powder than other sources of carbon, such as coal or graphite. It is porous, and as the mixture of materials is finely ground together, the other ingredients are packed into the pores of the charcoal and remain finely mixed. It has other volatile compounds in it besides carbon, which reduce the ignition temperature and speed the burning. The high carbon content leads to high temperatures when burning, which is essential in making the gases expand quickly.

Sulfur is widely available in nearly pure form near volcanic regions and hot springs. As volcanic gases escape from openings in the ground, the brim of the opening becomes covered in sulfur as the gases cool (hence the name *brimstone*). It was used for medicines, fumigation, and bleaching. In addition to sulfur mining, the Chinese had extracted sulfur from iron pyrite as early as the third century.

Sulfur ignites at a very low temperature, 190°C, and when finely powdered, can have a flash point as low at 166°C. Above about 444°C it becomes a gas. Along with the trace volatiles in the charcoal, sulfur speeds the burning of the final powder. The gases produced by the burning powder are carbon dioxide, carbon monoxide, and nitrogen, and at room temperature and normal atmospheric pressure, these gases would take up 380 times as much space as the original powder. But at the temperatures reached in the reaction, the gases expand to something closer to 3,000 times the original volume of powder. As this happens in twenty-five microseconds, a bang is heard.

Nitrates were likely the most difficult black powder ingredient to procure. While they exist in small quantities in animal dung, extraction would have been uneconomical. However, nitrates are a by-product of nitrifying bacteria (such as *nitrobacter* and *nitrosomonas*), which convert other nitrogen-containing molecules in compost heaps into nitrates. Over time (after about two years), a dung heap becomes richer in nitrates as the bacteria grow. The result is a much better fertilizer than animal dung alone, and farmers learned to tend the manure heaps by turning them to allow the bacteria to get oxygen, to add urine to the heaps as feedstock for the bacteria, and to build the piles on beds of clay and protect them from rain and sun. With all of these careful management processes, the result was rich enough in nitrates that crystals would form on surfaces.

Potassium nitrate decomposes when heated above about 550°C, releasing potassium nitrite and oxygen. In black powder at this temperature, the sulfur is already a gas and is well above its ignition temperature (190°C). The ignition temperature of charcoal (349°C) has also been exceeded at this point. Above about 790°C, the potassium nitrite then decomposes, releasing nitrogen and more oxygen to combine with the remaining sulfur and charcoal.

Early gunpowder may have been made from those unpurified crystals of potassium nitrate, sodium nitrate, ammonium nitrate, and strontium nitrate. But potassium nitrate can be purified from those dung-heap scrapings by making use of rudimentary chemical knowledge.

First, the nitrates are leached out of the nitrified manure by pouring water into it and collecting the nitrate-laden water the next day, which is then run through more nitrified manure, to get a higher concentration of nitrates. This extra concentration saves energy, as the water will later be boiled away.

The nitrate-laden water contains calcium nitrate, magnesium nitrate, sodium chloride, and potassium chloride, as well as the desired potassium nitrate. To convert the other nitrates to potassium nitrate, potassium hydroxide is added. This is done by adding wood ashes, which are rich in potassium hydroxide. The calcium and magnesium salts precipitate out of the solution and are filtered out.

The refining process continues by boiling the filtered liquid. Potassium nitrate is more soluble in boiling

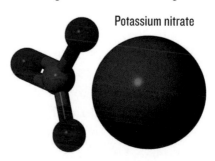

Potassium nitrate

water than common salt (sodium chloride) is. The salt gradually precipitates out, where it can be removed, and organic matter floats to the top, where it can be skimmed off.

When the liquid cools, the potassium nitrate crystallizes out and falls to the bottom, where it can be collected. The result contains 75% to 95% potassium nitrate. Further refining involves adding water to the crystals, adding protein (glue or blood) to coagulate any organic matter that can then be filtered out, and recrystallizing the potassium nitrate, stirring constantly as the mixture cools so that masses of crystals do not form (which would trap impurities between them).

Finally, the collected fine crystals are washed with a saturated solution of pure potassium nitrate (from prior refining). This cannot dissolve any more potassium nitrate, so the crystals do not dissolve, but it will wash away any remaining impurities and dissolve any salt.

As mentioned earlier, the first Chinese black powders were likely made with impure ingredients and without the optimum amount of nitrate oxidizer. As more experience with the manufacturing process evolved, recipes improved. One key discovery was that the burning happens faster if the mixture is very finely powdered. This is useful for making bombs and for use in small arms, but fast-burning fuel is not good for large cannons or rockets. To slow down the burning, the fine powder is mixed with small amounts of water and a binding agent (such as dextrin, a modified starch) and then forced through a sieve to make larger particles with less surface area. The ingredients thus remain finely mixed, as they were in the fine powder, which guarantees proper combustion, but the lower surface area reduces the combustion speed. At the same time, the spaces between the larger grains allowed the flame to set all the grains burning at once, so the pressure buildup started sooner and was more constant.

Burning black powder in air produces lots of nice smoke and flames, and it burns much faster than other materials do. This alone would have made it remarkable to early Chinese experimenters. But it is when black powder is confined that it produces the loud bang of the firecracker. It could be packed into bamboo tubes and thrown into a fire, creating an explosion.

The first paper firecrackers appeared around the tenth century, attributed to the Chinese inventor Li Tian. He is still celebrated in China every April 18, and parts of his workshop have been preserved by worshippers, giving evidence for the time line. Around the same time, the invention of the fuse allowed the user to light the firecracker and get away before it went off. Early fuses, made of straws or feather quills filled with black powder, were not very reliable. It was only much later (in the 1800s), when large amounts of black powder were used in mining, that more reliable fuses were developed.

During the Song dynasty in China (AD 960 to 1279), black powder was used in warfare as an incendiary. Since the recipes used then were low in nitrates, they burned rather than exploded. Black powder, wrapped in paper and attached to the shafts of arrows and equipped with a fuse, burned hotter than the burning pitch used earlier and was harder to put out, since it had its own oxygen source. Hollow metal balls with holes in them were filled with black powder and lofted at enemies from catapults. Again, the very hot flames shooting out the holes ignited whatever flammable material they encountered. Sometime around 1100 the emperor Jen stopped all export of sulfur and saltpeter, and made a state monopoly out of the manufacture of black powder to prevent its use by enemy armies.

The Song dynasty (at least the northern part) fell to the Jin invaders from Korea when the capital Kaifeng was attacked. During that battle, in the year 1126, the first exploding bombs were used against the invaders. Because the low nitrate content of the powder could not burn fast enough to burst strong container walls, they were not much more than big firecrackers made of bamboo or paper and produced mostly noise and smoke, although the bombs contained as much as four pounds of powder and were probably effective incendiaries as well. But despite this new technology, the battle was lost in 1127.

By 1150 the invaders (now the Jin dynasty) were manufacturing saltpeter in manure heaps designed for that purpose and manufacturing black powder for weapons. They perfected the recipes, adding more nitrate until the powder could burst iron bomb casings. By holding up against more pressure, the bombs could deliver much more power than the earlier paper bombs could. When, in 1231, the invading Mongols attacked Kaifeng, these bombs were used to defend the city. The explosions charred areas scores of yards square and could be heard over 30 miles away. The shrapnel could pierce the Mongol iron armor, and bombs lowered on chains from the city walls blew the attackers to pieces.

Eventually, however, the Mongols under Genghis Khan prevailed and the Jin dynasty fell. The Mongols, with their highly effective cavalry tactics, brought the new weapons to later conflicts. By 1274, under Kublai Khan, the grandson of Genghis, they attacked the remaining Southern Song dynasty, adding the remainder of China to their expanding empire.

For the duration of the Mongol Empire, bomb-making technology continued to improve. Larger bombs, bombs with added shrapnel, and bombs with shrapnel mixed with noxious chemicals and waste designed to maim, blind, and infect the enemy were created and deployed. But this was also the era of the rocket.

The first rockets were likely the arrows discussed earlier, equipped with black powder in paper tubes attached to the shaft. A relatively small change to that design resulted in a tube that generated extra thrust for the arrow as the powder burned, exiting through the hole made for the fuse.

Fire lances, which were handheld tubes filled with black powder and ignited, had been used for some time already as frontline weapons, spewing fire for several feet in front of the operator. Think of these as an early flamethrower. The flames would persist on the order of five minutes or so. An operator of such a weapon might feel a force pushing back, but it would not be much unless there was some constriction (a nozzle) where the flames left the tube. This would allow pressure to build up in the tube, propelling the gases forward at a higher speed. Fire lances were used by the Jin defending Kaifeng from the Mongols in 1232 and may have been in use as early as 1000.

As early as 1256, when the Mongols attacked the Assassins, the use of rocket-assisted arbalest arrows allowed ranges of up to two miles. A type of large crossbow, the arbalest likely provided much of the propulsive power, but to reach that distance, the rocket would have certainly helped. With a delayed fuse, a rocket could take over the propulsion before the arrow pointed down, adding to the range and possibly to the penetrating power on impact. It would not have had to be powerful enough to launch itself.

When a rocket is self-launched, its initial velocity is low, and this greatly reduces the accuracy. Starting with an arrow launched from a powerful arbalest, the rocket would ignite when the aerodynamics of the speeding arrow had already stabilized it, making it an effective weapon. And there would obviously be less powder needed in the rocket. Without the rocket assist, arbalests of the time had a range of about a quarter of a mile.

Unassisted by the speed and aerodynamics of an arrow, a tube filled with powder and closed at one end would propel itself along the ground in a random fashion, unable to aim. Such devices, known as "ground rats," were in use sometime after 1224 and provided amusement in the court of the emperor.

One record of rockets being used as signaling devices dates to 1272, during the Mongol siege of Xiangyang. Defenders had quietly left the city in boats in the night to try to bring in supplies. Seeing masts in the dark, they signaled what they thought were Song ships come to help. The ships were instead manned by Mongols, who saw the flares and captured the boats and crews, leaving the city cut off from supplies. These signal rockets may have been the first rockets that did not need the assistance of a launching bow, as accuracy would not be as important for this use.

Rockets continued to improve in subsequent years. Technicians learned to increase the surface area of the burning front by making the powder charge hollow. The powder thus formed its own tube inside the containing tube. Nozzles were made to increase the pressure, so the gases would escape at higher speed. This greatly increased the propulsive force. As with the arrows before them, rockets could carry explosive charges. Adding a second stage, so that one rocket lifted another one, gave them still greater range.

Not much has survived regarding the methods by which the early makers of black powder practiced their craft. But there are more recent documents, such as one from the American Civil War, where the practice of making saltpeter (potassium nitrate) for the Confederate war effort was laid out in a pamphlet encouraging farmers to produce it.

The Northern blockade was quite effective in cutting off supplies to the Confederacy. As early as 1862, barely a year into the four-year war, it was recognized that supplies of critical war materials, such as saltpeter, might run out if hostilities continued longer than Confederate planners had expected. Since the production of saltpeter takes eighteen to twenty-four months, getting farmers to begin production was prudent.

The task fell to Joseph LeConte, a professor of chemistry and geology in South Carolina. In fourteen pages, in *Instructions for the Manufacture of Saltpetre*, he discusses the current state of the art in making saltpeter. He begins:

By the request of the Chief of the Department of the Military, under authority of the Executive Council, I have been induced to publish, for the instruction of planters and manufacturers, a very succinct account of the most approved methods of manufacturing saltpetre. In doing so, I shall aim only at brevity and clearness.

The general conditions necessary to the formation of saltpetre are: 1st, the presence of decaying organic matter, animal or vegetable, especially the former; 2d, an alkaline or earthy base, as potash or lime; 3d, sufficient moisture; 4th, free exposure to the oxygen of the air; and 5th, shelter from sun and rain. . . .

By these means, if diligently used in all parts of the State, it is hoped that an immediate and not inconsiderable amount of saltpetre may be obtained. It is not believed, however, that the supply thus obtained will be sufficient for the exigencies of the war. It is very important, therefore, that steps should be taken to insure a sufficient and permanent supply of this invaluable article. This can only be done by means of nitre-beds. I proceed, then, to give a very brief account of the method of making these.

The most important prerequisite in the construction of nitre-beds in such manner as to yield nitre in the shortest possible time, is a good supply of thoroughly rotted manure of the richest kind, in the condition usually called mould, or black earth. It is believed that in every vicinity a considerable supply of such manure may be found, either ready prepared by nature, or by the farmer and gardener for agricultural and horticultural purposes.

To make the bed, a floor is prepared of clay, well rammed, so as to be impervious to water. An intimate mixture is then made of rotted manure, old mortar coarsely ground, or wood ashes (leached ashes will do), together with leaves, straw, small twigs, branches, &c. to give porosity to the mass, and a considerable quantity of common earth, if this has not been sufficiently added in the original manure-heap. The mixture is thrown somewhat lightly on the clay floor, so as to form a porous heap four or five feet high, six or seven wide, and fifteen feet long. The whole is then covered by a rough shed to protect from weather, and perhaps protected on the sides in some degree from winds.

The heap is watered every week with the richest kinds of liquid manure, such as urine, dung-water, water of privies, cess-pools, drains, &c. The quantity of liquid should be such as to keep the heap always moist, but not wet. Drains, also, should be so constructed as to conduct any superfluous liquid to a tank, where it is preserved and used in watering the heaps.

The materials are turned over to a depth of five or six inches every week, and the whole heap turned over every month. This is not always done, but it hastens very much the process of nitrification. During the last few months of the process, no more urine, nor liquid manure of any kind, must be used, but the heaps must be kept moist by water only. The reason of this is, that undecomposed organic matter interferes with the separation of the nitre from the ley. As the heap ripens, the nitre is brought to the surface by evaporation, and appears as a whitish efflorescence, detectible by the taste.[1]

I imagine it was not that common for farmers to routinely check the taste of their manure heaps, but LeConte seems unconcerned with this issue.

---

[1] This work is the property of the University of North Carolina at Chapel Hill.

When this efflorescence appears, the surface of the heap is removed, to the depth of two or three inches, and put aside under shelter, and kept moist with water. The nitre contained is thus considerably increased. When the whitish crust again appears, it is again removed until a quantity sufficient for leaching is obtained. The small mound which is thus left is usually used as the nucleus of a new heap. By this method it is believed that an abundant supply of nitrified earth, in a condition fit for leaching, may be obtained by autumn or early winter.

LeConte describes four methods of preparing the nitre beds. The first is the French method, where the compost is mixed with mortar and ashes in layers, so the alkali in the mortar and ash can work on the nitrates as the bacteria are producing it (of course he knows nothing about bacteria at this time).

The second method is the Prussian method:

Five parts of black earth and one of spent ashes or broken mortar are mixed with barley straw, to make the mass porous. The mixture is then made into heaps six feet high and fifteen feet long with one side perpendicular (and hence called walls), and the opposite side sloping regularly by a series of terraces or steps.

Straight sticks are generally introduced, and withdrawn when the mass is sufficiently firm. By this means air and water are introduced into the interior of the mass. The heap is lightly thatched with straw, to protect from sun and rain. The whole is frequently watered with urine and dung-water.

The perpendicular side being turned in the direction of the prevailing winds, the evaporation is most rapid on that side. The liquid with which the heap is watered is drawn by capillarity and evaporation to this side, carrying the nitre with it, and the latter effloresces there as a whitish crust. The perpendicular wall is shaved off two or three inches deep as often as the whitish incrustation appears, and the material thus removed is kept for leaching.

The leached earth, mixed with a little fresh mould, is thrown back on the sloping side of the heap, and distributed so as to retain the original form of the heap. Thus the heaps slowly change their position, but retain their forms. This method yields results in about a year—probably in our climate in eight months.

The Swedish method is simpler, just mixing everything in a heap and turning it:

Every Swede pays a portion of his tax in nitre. This salt is therefore prepared by almost every one on a small scale. The Swedish method does not differ in any essential respect from those I have already described.

First a clay floor; upon this is placed a mixture of earth, mould, spent ashes, animal and vegetable refuse of all kinds. Small twig branches, straw and leaves are added, to make the mass porous; a light covering, to protect from weather, frequent watering with urine or dung-water, and turning over every week or two.

The process is precisely the same as the French, except that the process of preparation and nitrification are not separated. I only mention it to show that nitre may be made by every one on a small scale. By this method the beds are ripe in two years—perhaps in less time in this country.

The last method is the Swiss method, where a stable is built on a hill, collecting the animal droppings in a pit:

The method practiced by the small farmers in Switzerland is very simple, requires little or no care, and is admirably adapted to the hilly portions of our State.

A stable with a board floor is built on the slope of a hill (a northern slope is best), with one end resting on the ground, while the other is elevated, several feet, thus allowing the air to circulate freely below.

Beneath the stable a pit, two or three feet deep, and conforming to the slope of the hill, is dug and filled with porous sand, mixed with ashes or old mortar.

The urine of the animals is absorbed by the porous sand, becomes nitrified, and is fit for leaching in about two years. The exhausted earth is returned to the pit, to undergo the same process again.

This leached earth induces nitrification much more rapidly than fresh earth; so that after the first crop the earth may be leached regularly every year.

A moderate-sized stable yields with every leaching about one thousand pounds of saltpetre.

Once the nitre beds have produced the nitrates, they must be extracted from the mess:

When the process of nitrification is complete, the earth of the heaps must be leached. Manufacturers are accustomed to judge roughly of the amount of nitre in any earth by the taste. A more accurate method is by leaching a small quantity of the earth, and boiling to dryness, and weighing the salt.

There is much diversity of opinion as to the percentage of nitre necessary to render its extraction profitable. The best writers on this subject vary in their estimates from fifteen pounds to sixty pounds of salt per cubic yard of nitrified earth. The high price of nitre with us at present would make a smaller percentage profitable. This point, however, will soon be determined by the enterprising manufacturer.

In the process of leaching, in order to save fuel, we must strive to get as strong a solution as possible, and at the same time to extract all or nearly all the nitre. These two objects can only be attained by repeated leachings of the same earth, the ley thus obtained being used on fresh earth until the strength of the ley is sufficient.

A quantity of nitrified earth is thrown into a vat, or ash-tub, or barrel, or hogshead with an aperture below, closely stopped and covered lightly with straw. Water is added, about half as much in volume as the earth. After stirring, this is allowed to remain twelve hours. Upon opening the bung, about half the water runs through containing, of course, one-half the nitre. Pure water, in quantity half as much as first used, is again poured on, and after a few moments run through. This will contain one-half the remaining nitre, and therefore one-fourth of the original quantity. Thus the leys of successive leachings become weaker and weaker, until, after the sixth leaching, the earth is considered as sufficiently exhausted. The exhausted earth is thrown back on the nitre-beds, or else mixed with black earth to form new beds. The leys thus obtained are used upon fresh earth until the solution is of sufficient density to bear an egg. It then contains about a pound of salt to a gallon of liquid.

The clever way of determining the specific gravity (density) of the solution on a farm without special equipment is notable: just see if the solution will float an egg.

The next step is to convert all the nitrates into potassium nitrate. Nitrates of lime and magnesia refer to calcium and magnesium nitrates. The calcium and magnesium take up the carbonate and hydroxide from the ashes and precipitate out:

The ley thus obtained contains, besides nitrate of potash (nitre), also nitrate of lime and magnesia, and chlorides of sodium and potassium. The object of the next process is to convert all other nitrates into nitrate of potash. This is done by adding wood ashes. The potash of the ashes takes all the nitric acid of the other nitrates forming the nitrate of potash (nitre), and the lime and magnesia are precipitated as an insoluble sediment. Sometimes the ashes is mixed with the nitrified earth and leached together, sometimes the saltpetre ley is filtered through wood ashes, sometimes the ley of ashes is added to the saltpetre ley. In either case the result is precisely the same.

The ley thus converted is then poured off from the precipitate, into copper or iron boilers. It still contains common salt (chloride of sodium) in considerable, and some other impurities in smaller, quantities. It is a peculiarity of nitre, that it is much more soluble than common salt in boiling water, but much less soluble in cold water. As the boiling proceeds, therefore, and the solution becomes more concentrated, the common salt is, most of it, precipitated in small crystals, as a sandy sediment, and may be raked out.

Much organic matter rises as scum, and must also be removed. When the concentration has reached almost the point of saturation, the boiler must be allowed to cool. This is known by letting fall a drop of the boiling liquid upon a cold metallic surface; if it quickly crystallizes, it is time to stop the boiling.

It is now poured into large receivers and left to cool. As the ley cools, nearly the whole of the nitre separates in the form of crystals, which sink to the bottom. These are then removed, drained by throwing them in baskets, and dried by gentle beat. The mother-liquor is either thrown back into the boilers, or else used in watering the heaps.

The product thus obtained is the crude saltpetre of commerce. It still contains fifteen to twenty-five per cent of impurities, principally common salt (chloride of sodium), chloride of potassium and organic matter. In this impure form it is usually brought to market.

LeConte could have stopped there, as his job is done—he has instructed farmers in how to bring potassium nitrate crystals to market, where refiners can process them into a form pure enough for gunpowder. Instead, he continues, describing the final refining process:

One hundred gallons of water is poured into a boiler, and crude saltpetre added from time to time, while the liquid is heating, until four thousand pounds are introduced. This will make a saturated solution of nitre.

The scum brought up by boiling must be removed, and the undissolved common salt scraped out. About sixty gallons cold water is now added gradually, so as not to cool the liquid too suddenly. From one to one and a-half pounds of glue, dissolved in hot water, is added, with stirring. Blood is sometimes used instead of glue. The glue seizes upon the organic matter, and they rise together as scum, which is removed.

Continue the boiling until the liquid is clear. The liquid is then suffered to cool to one hundred and ninety-four degrees, and then carefully ladled out into the crystallizers. These are large shallow vats, with the bottom sloping gently to the middle. In these the cooling is completed, with constant stirring. In the process of cooling nearly the whole of the nitre is deposited in very fine, needle-like crystals, which, as they deposit, are removed and drained.

In this condition it is called saltpetre flour. The object of the constant stirring is to prevent the aggregation of the crystals into masses, from which it is difficult to remove the adhering mother-liquor.

The saltpetre flour is then washed of all adhering mother-liquor. For this purpose it is thrown into a box with a double bottom; the lower bottom with an aperture closely plugged, and the false bottom finely perforated. By means of a watering pot a saturated solution of pure nitre is added, in quantity sufficient to moisten thoroughly the whole mass.

After remaining two or three hours to drain, the plug is removed and the solution run out. This is sometimes repeated several times. The saturated solution of nitre cannot, of course, dissolve any more nitre, but dissolves freely the impurities present in the adhering mother-liquor.

Last of all, a small quantity of pure water—only about one pound to fifty-three pounds of the nitre to be washed—is added in the same manner, and run off at the end of two hours.

The nitre is now dried by gentle heat and constant stirring, and may be considered quite pure, and fit for the manufacture of gunpowder.

After a section on how to analyze the resulting crystals for purity, LeConte concludes with some words for those who might think this is all too much trouble:

In conclusion, a word by way of encouragement to manufacturers in undertaking this work.

It will be seen that under the most favorable circumstances saltpetre cannot be made in any considerable quantity in less than six or eight months, and that if we commence now the preliminary process of preparing black earth, so as to insure a sufficient and permanent supply, results cannot be expected under eighteen months or two years.

Let no one be discouraged by this fact, under the idea that the war may not last so long, and all their work may be thrown away. There is every prospect now of the war continuing at least several years, and of our being thrown entirely on our own resources for war materials. Besides, even if the war should be discontinued, the work is by no means lost.

The method of preparing and making saltpetre-beds is precisely the most approved method of making the best manure, and all the labor and pains necessary for the preparation of black earth, and the construction of saltpetre-beds, and which I hope to induce my fellow-countrymen to undertake under the noble impulse of patriotism, ought to be annually undertaken by every planter, under the lower impulse of a wise self-interest, and would be amply rewarded in the increased production of field crops.

And what became of Joseph LeConte? He was thoroughly disgusted by the loss of the war and by Reconstruction, claiming that the "sudden enfranchisement of the negro without qualification was the greatest political crime ever perpetrated by any people." Apparently, the irony of this statement in the face of slavery was lost on him. He fled the conditions he so deplored and moved to Berkeley, California, to become a professor of geology there. He became friends with John Muir in Yosemite Valley and cofounded the Sierra Club.

BLACK POWDER GUNS

As the quality of the black powder improved (largely through the use of more potassium nitrate with higher purity), the fire lance improved along with it. The new powder shot the flames out with more force. Bits of metal and broken pottery could be added, and this shrapnel would shoot out toward the enemy with enough velocity to cause injury.

With the advent of metal tubes instead of bamboo, stronger powder, and projectiles, the fire lances became too heavy to hold by hand, prompting the addition of wheeled carts to move them around more easily. These "erupters," as they were called, defended fortifications. Some shot multiple arrows from the tubes, and some shot metal balls. As the projectiles grew to fill the mouth of the tube, the force of ejection increased. When the powder was no longer used for its flame, and instead was used only to propel projectiles, the true gun had arrived.

Handguns made with metal barrels were in use by the late 1200s. One gun, found in 2004, is inscribed with the date 1271, five years before the Mongols overran the Southern Song. It was made of copper and shaped like an elongated vase, bulbous in the rear where the powder exploded. Weighing a little over three pounds, it was about 13 inches long and had a 1-inch bore.

Another gun, excavated at an archaeological dig in Heilongjiang, dates to 1288. Like the earlier weapon, it also has a 1-inch bore, but the bulbous rear had already become a simple tube. It was made of cast iron, with reinforcing hoops of iron at intervals along the barrel, reminiscent of the nodes of a stalk of bamboo.

By 1276 Kublai Khan and his Mongols, now known as the Yuan dynasty, had conquered the Song dynasty in the south, once again uniting all of China under one rule. The Mongol Empire stretched from China all the way west to Russia and the Middle East. While Mongol cavalry, tactics, and sheer numbers were responsible for the military successes, guns and gunpowder played a role in several important battles. The Song used explosive land mines in 1277 against the Mongol invaders, who nonetheless prevailed, using that technology later. The trigger was a pull of a cord that struck flint against steel to spark the explosion.

While the hand cannons used by Chinese fighting under Mongol rule in Manchuria in 1288 had the hooped cylindrical form described earlier, some larger artillery pieces still used the more bulbous shape as late as 1350. When better metallurgical techniques allowed higher strength materials, the cylindrical form made its way into larger guns.

The Red Turban Rebellion that eventually led to the toppling of the Mongol Yuan dynasty in the mid-1300s used cannons that fired arrows in battles on land as well as at sea. The Ming dynasty that followed used cannons in battles in the south in the late 1300s, but it wasn't until later that they were used in the north, particularly in 1414, when used against the western Mongol Oirats, reportedly killing several hundred. The impressive noise of the cannon is thought to have helped in the victory.

Much later, in 1593 at the siege of Pyongyang, the Ming used cannons effectively against a Japanese army that matched them in size but lacked the firepower. The Japanese were defeated in a single day, due for the most part to the superior weaponry.

The Mongols had brought guns and gunpowder to the Middle East, and Muslim armies brought them into Spain in the early 1200s. Cannons played a part in the siege of Seville, helping Christian armies to take the city in 1248.

Hand cannons found use as early as 1260 in the battles between the Mongols and the Muslim Mamluks, in Ain Jalut, just north of Jerusalem. The Egyptian Mamluks were using hand cannons to frighten Mongol horses and cause disorder.

In 1274 the sultan Abu Yaqub Yusuf used cannons in the siege of Sijilmasa in Morocco.

Farther north, gunpowder was known in England as early as 1216, when Roger Bacon first wrote about it. In 1248 he published a formula for it and described its military uses. Cannons were used to fire arrows and small metal balls (grapeshot) in the mid-1300s in England and France during the Hundred Years' War. In the later 1300s, wheeled cannons started to appear.

Small cannons and firearms were cast from bronze or iron. The ability to cast larger pieces, however, took much longer to develop. In the meantime, the art of the blacksmith was put to use, and cannons were made from wrought iron. A large wooden pole was used as a core, and bars of red-hot iron were pounded together to weld them into a single piece around the pole. The next step used technology borrowed from the makers of wine barrels. Hoops of iron were brought to red heat to make them expand. They were then slipped over the wrought iron bars and allowed to cool and shrink around them, reinforcing the bars against the pressure of the exploding powder. It was this technology that gives the name to this part of the gun: the barrel.

The powder was put into a separate chamber called the powder chamber. Once the powder was poured into this upright chamber, a wooden bung was tamped into place on top, and the whole thing turned horizontally and attached to the end of the barrel. The powder chamber had a smaller bore than the cannon barrel, so the walls could be thicker to withstand the explosion. A red-hot poker was brought to the touchhole to set it off.

The balls for these cannons were made from stone, shaped into rough spheres by stonemasons. One of the big Flemish bombards (built in 1450) weighed 18 tons and fired stone balls two feet in diameter.

Cannons were used to defend Constantinople in 1396, but later, in 1422, the Ottoman attackers returned with improved large cannons of their own that could send stone balls weighing hundreds of pounds to distances over a mile. These huge bombards—one was 27 feet long—took hours to reload and were dragged to battle by teams of sixty oxen and four hundred men.

These huge guns were cast from copper, not made of wrought iron. The wrought iron guns could not stand up to the power of the new powders. The casting was done at the site of the siege, as the guns were enormous. (One had a barrel 26 feet long and shot a stone weighing over a thousand pounds.) Fifty oxen were needed to move the gun, and the crew that fired it numbered seven hundred men. The first stone shot from the gun went farther than a mile and buried itself 6 feet into the ground.

Cast cannons could no longer be breech loaded. The powder chamber was now part of the whole gun, cast in the same pour as the barrel. The guns were now loaded from the muzzle end.

Cannons changed the way Europeans built fortifications. Old-fashioned castle walls were vulnerable to cannon fire, the flat stone walls taking the impact directly, the stone shattering. Replacing them were star formations, with sloping sides made of earth, sometimes faced with stone or brick (which did not shatter like stone), to deflect the cannonballs before they could get to the actual walls. The star formation allowed defenders to fire on anyone trying to dig under the walls, as there was nowhere to hide.

Field artillery continued to improve by the late 1400s. Cannons on wheels allowed mobile forces more firepower and more options in its deployment. The barrels of the cannons were cast with side projections near the center of mass, called trunnions, so changes in elevation were easier and faster to make. No longer just used for battering at walls, the wheeled artillery was now used against troops, cavalry, and enemy cannons. Cannons became smaller as improved metallurgy and better powder allowed higher velocities. The lighter cannons were easier to move around and took less time to load.

Despite these improvements, field cannons still moved at a walking pace at best, and much of warfare was done with small arms and pikes. Small firearms such as the hand cannons, which were fired by touching a flame or hot coal to the powder in a touchhole, made way for the arquebus, a shoulder-fired long gun that was fired by moving a lever to lower a lit piece of cord (a "match") onto the powder. This allowed the user to keep his eyes on the target and both hands on the weapon. Such "matchlock" guns had a small cup, called a flash pan, that held the priming powder on the side of the gun, connecting to the touchhole, which was no longer on the top of the barrel. The barrel was flared at the muzzle end to make it easier to muzzle-load the powder and ball. The mechanism was called a "lock" because it resembled the lock on the door of a house. The phrase "lock, stock, and barrel" comes from the three main parts of a gun.

With the arquebus came the invention of volley fire, where once the front rank of guns had fired, a second rank took their place, and then a third, by which time the first rank had been able to reload. With this innovation, the arquebus became the main tool for armies, instead of being merely a support piece. By the late 1400s and early 1500s they were used in large numbers on the battlefield, and Hernán Cortés brought some with him to Mexico during his conquest of the Aztecs.

Matchlock weapons had many drawbacks. In wet weather the match often went out. Carrying a lit match when loading a firearm or handling gunpowder was obviously dangerous. The matches were visible at night, and the odor of the burning match carried far downwind, making surprises less likely. On top of that, the match had to stay lit for long periods, which required a large supply of the saltpeter impregnated cord.

The next improvement was the wheel lock, which was invented around the year 1500. Some think it was invented by Leonardo da Vinci, but there is evidence it was previously invented by an unknown German. A pull of the trigger caused a steel wheel to spin very quickly, rubbing against a piece of iron pyrite to produce a spray of hot sparks. These would hit the primer powder in the flash pan and fire the weapon. Such self-igniting guns overcame many of the disadvantages of the matchlock. The flash pan now also had a cover (slid out of place by the trigger mechanism) that kept the powder in place and dry.

Wheel lock weapons were complex and difficult to manufacture. Later mechanisms such as the snaplock and flintlock used a flint that hit a piece of steel, sending white-hot steel fragments into the flash pan. These were cheap to manufacture and generally replaced wheel lock weapons everywhere. The steel part the flint hit, called the frizzen, is an integral part of the flash pan cover. As the sparks are created, the frizzen is kicked out of the way, opening up the flash pan to receive the sparks.

By the mid-1500s gunpowder manufacturing had improved considerably. The nitrate percentage was improved to 75%, and the powder was *corned*. Corning the powder meant forming it into larger pieces by extruding it through a sieve. This allowed the flame front to quickly move through the open spaces between the grains, causing the entire charge to ignite at once, instead of having to burn through packed powder from one end to the other. Since the powder was still finely ground before corning, the oxidizer and fuel were still in very close contact and the powder burned very fast.

By the late 1500s cast-iron had replaced stone balls as projectiles, and cannon bores were standardized, so that mass produced balls would fit the artillery pieces closely, ensuring a better seal. The improved gunpowder and metallurgy allowed smaller artillery pieces to be very effective as the speed and range of the projectiles increased.

While the Chinese had been throwing explosive bombs for quite some time, using trebuchets and catapults, cannons that could do this effectively began to appear in Europe in the mid-1300s. Exploding shells were used in battles in Italy in 1376 and 1421. These shells had fuses that were loaded facing the main powder charge in the cannon, and the timing of the explosion was an issue. If the fuse was too short, the shell would burst in air prematurely, or burst inside the cannon. A fuse that was too long would allow the enemy time to run from the bomb. To ensure that the fuse was always facing the powder charge, a wooden *sabot* (shoe) was attached to the spherical projectile.

Shells were particularly effective when combined with the mortar, a short cannon used for sending shells upward over walls, hills, or other fortifications.

For the most part, cannons continued to shoot solid spherical balls well into the 1800s. Solid shot was easy to manufacture and had few of the technical problems that exploding shells had.

In 1784 Henry Shrapnel invented a shell that contained a mixture of musket balls and black powder connected to a time fuse. The shell would explode in midsair and release the musket balls, which, because of the initial velocity of the shell, carried deadly force over a wide range onto unarmored infantry.

Rifled cannons began to appear as early as 1840. Rifled small arms were common by then, but the technology to rifle large barrels took longer to develop. Rifling spins the projectile, giving it better accuracy, but it also brings with it other advantages. In order for the rifled barrel to spin the bullet or shell, the projectile must fill the barrel. Muzzle loaded smoothbore cannons fired shot that was smaller than the cannon bore, to make it easier to load. Various schemes to close the gap between the bullet and the bore included expanding copper dishes behind the shell and lead rings around the rear of the shell that engaged the rifling. These trapped the gas and allowed as much as twice the amount of gas to actually move the projectile.

Muzzle loaded rifled barrels, whether small arms or cannons, were difficult to load, as the bullet or shell had to be

Black powder flask with measuring spout

rammed past the rifling. Smoothbore muskets were the common tool of the military long after the rifle was invented, because it was much faster to load. That began to change in the mid-1800s. In 1823 Captain John Norton of the British 34th Regiment saw natives in South India using blowguns. He saw how the pith behind the blowgun dart expanded to fill the tube when pushed by the user's breath.

In 1832 Norton invented a cylindrical bullet with a cone at the front, with a hollow base that would expand when the powder exploded. This sealed it against the bore of the gun, so that no gas escaped around the bullet. The result was much faster acceleration and a higher muzzle velocity. In 1836 London gunsmith William Greener improved the design by making the bullet oval with a flat base, with a hole drilled into it. A conical plug with a wooden base fit into the hole. When the powder exploded, the cone pushed into the bullet, widening it to form a good seal.

In 1849 two French army captains, Claude-Étienne Miniè and Henri-Gustave Delvigne created a soft-lead bullet with a hollow base, surrounded by several grooves to hold grease. The bullet was cylindrical and longer than a musket ball but still narrow enough to easily drop down the barrel of a rifle.

The British used this new bullet in the Crimean War to devastating effect against the Russian soldiers, who were equipped with older smoothbore muskets. The longer range and greater accuracy, combined with the deadly effect the new faster bullets had on flesh and bone, made newspaper headlines in the *London Times*. They called the Miniè equipped rifle "the king of weapons."

The United States had military observers at the Crimean War battles, and the new bullets impressed them as much as they had the *Times*. The armorer at the US arsenal at Harpers Ferry, Virginia, John Burton, adapted the new bullet for mass production, getting rid of the drilled hole and conical base with its wooden disk, and replacing those design features with a simple hollow base.

The US secretary of war at the time was Jefferson Davis (later to be the president of the Confederate States). He adopted the new bullet for the US Army in two sizes, a .69-caliber and a .58-caliber.

When the American Civil War broke out, both sides were still mostly using the old smoothbore muskets. As the war progressed, however, rifled muskets became much more common. The smoothbore weapons had an effective range of about 50 yards. The bullets could travel as far as 200 yards, but the accuracy was so poor that they were only effective at the shorter range. Rifles with the Miniè ball were effective at 300 yards, and the bullets flew as far as half a mile. The huge numbers of casualties in the Civil War can be traced to the more effective rifles and the resistance of military leaders to adjust their tactics to the new realities.

Rifled cannons came into much more common use when breech loaded cannons were invented in the mid-1800s.

Corned powder was not the last improvement made. In the 1800s powder makers discovered that compressing the damp powder at high pressures (over 1,200 pounds per square inch) made it considerably denser. Denser explosives are more powerful, not only packing more power (and powder) into the same volume but also getting the fuel and the oxidizer in even closer proximity. The dense hard cakes could no longer be forced through a sieve but had to be broken first into pieces. The pieces were sieved and any too large were broken further, while any too small were re-moistened and compressed into new cakes. The pieces were then tumbled with graphite in a drum. This gave them a polished coating of graphite, which both kept out moisture and conducted electricity, making electric sparks less of a hazard.

3

FULMINATING COMPOUNDS

Medieval alchemists knew that gold could be dissolved in a mixture of nitric and hydrochloric acids, known as aqua regia. One of the goals of alchemy, after all, was the production of gold from cheaper materials, and divining the properties of gold was an obvious first step. They also knew that a basic substance like lye or ammonia could neutralize acids. Thus, it was almost inevitable that one of them would eventually dissolve gold in aqua regia and then try to neutralize the acid with ammonia or potassium carbonate.

Imagine their surprise when it exploded in their faces.

Bang.

In 1585 an alchemist in Germany named Sebald Schwartzer wrote out a formula for making what was then known as *aurum fulminans*, a compound today called fulminating gold. It calls for adding eight parts of gold to concentrated nitric acid, warming it in a bath of sand, then adding eight parts ammonium chloride. This dissolves the gold—the nitric acid allows the chloride ions to combine with the gold to make hydrogen tetrachloroaurate(III). By making the aqua regia with ammonium chloride instead of hydrogen chloride, the resulting liquid contained quite a few ammonium ions.

The recipe then calls for adding a mixture of iron and copper sulfate solutions to the mix to get a precipitate, which is filtered out, washed, and concentrated by evaporation into a thick solution. Finally, concentrated potassium carbonate (a strong alkali) is added carefully, so that it does not cause heavy foaming and overflow, or worse, explode. The precipitate is purple crystals of fulminating gold. The recipe says the whole process takes four days but no more than five days.

The word *fulminate* means to explode violently. (It comes from the Latin word for lightning.) First applied to this compound of gold, it was later used to describe other compounds that exploded violently, such as compounds of silver and mercury. The latter compound gives name to a class of compounds called fulminates. Fulminating gold and fulminating silver are not fulminates, they just fulminate.

Chemists tried to analyze fulminating gold for centuries. It seems that no two batches of it are quite the same, and even in a single batch, there are compounds with different properties. The conclusion at this time is that each gold atom is connected to a nitrogen atom, and that oxygen atoms and chlorine atoms connect these into a large, noncrystalline, three-dimensional polymer.

The explosive nature of fulminating gold was noteworthy, and chemists experimented with it, but it was clearly too expensive to use in weapons or mining. However, it remains the first high-explosive compound discovered. A high explosive is a chemical compound where the oxidizer and the fuel are combined in the same molecule. You may recall from the discussion of the manufacture of gunpowder that many of the improvements were due to the finer and finer mixing of the nitrates with the sulfur and charcoal. The closer the fuel is to the oxidizer, the faster the reaction can happen.

Black powder merely burns if it is not contained. In high explosives, the reaction takes place faster than the speed of sound. This means that the air cannot move out of the way fast enough to let the exploding gases out, and the air itself (and the inertia of the initial ingredients) acts like a container. There is a sharp report as the material disintegrates into gas so fast that it generates very high pressures.

In 1786 the prolific French chemist Claude-Louis Berthollet added ammonia to precipitated silver oxide and produced fulminating silver. It was highly sensitive, exploding easily by touch or slight heating. It was also quite powerful, and, like fulminating gold, it was a high explosive.

Berthollet is known for many chemical discoveries, demonstrating the bleaching effects of chlorine gas, creating sodium hypochlorite (the bleaching agent in Clorox bleach) and (of particular interest here) potassium chlorate, a strong oxidizing agent that became known as Berthollet's Salt.

In 1789 another French chemist (and contemporary of Berthollet), Antoine François, comte de Fourcroy, described another fulminating mixture, called fulminating powder. It was a mixture of three parts potassium nitrate, two parts potassium carbonate, and one part sulfur. The ingredients were ground together in a hot marble mortar with a wooden pestle, and the powder then placed on a metal ladle and warmed until the sulfur melted. After a short while, the mixture then exploded, denting or perforating the iron ladle. In further experiments, he noted that a mixture of one part potassium sulfide to two parts potassium nitrate "fulminates more rapidly," leading him to conclude that the first mixture created potassium sulfide shortly before detonating.

Fourcroy also described another fulminating mixture, made by replacing the potassium nitrate in gunpowder with Berthollet's new discovery, potassium chlorate. This new mixture would explode when hit with a hammer.

In 1797 Pierre Bayen published Volume I of his *Opuscules Chimiques*, where he describes fulminating mixtures of mercury compounds with sulfur, which explode when heated.

Around this time, Luigi Brugnatelli (who later would become famous for gold electroplating) experimented with nitrates of gold, silver, and mercury, and noted that compressing them with sulfur or phosphorus could produce detonations. Belgian chemist Jean-Baptiste Van Mons also worked with oxides of noble metals combined with sulfur or phosphorous to make explosive mixtures, and notes that such mixtures produce a more uniform explosive effect than the mixtures containing potassium chlorate.

Edward Charles Howard was the son of a Catholic wine merchant in Sheffield, England. His parents believed he could not get a proper Catholic education in Protestant England, so at the age of nine he was sent to study in France at the Catholic English College in Douay. In 1788, when he reached fourteen, he returned to England, despite having finished only half of his schooling. His father had died six months earlier, and the troubles about to lead to the French Revolution were already apparent. Over the next twelve years he became a highly skilled chemist and was elected to be a Fellow of the Royal Society. His election was supported by his third cousin, the Eleventh Duke of Norfolk, and several other prominent scientists, including Peter Woulfe, who had discovered picric acid a few years earlier.

In 1800, at the age of twenty-six, Howard published an account of his experiments with fulminating mercury in the *Philosophical Transactions* of the Royal Society. He describes what is now called mercury fulminate this way:

> The mercurial preparations which fulminate, when mixed with sulphur, and gradually exposed to a gentle heat, are well known to chemists: they were discovered, and have been fully described, by Mr. Bayen.
>
> MM. Brugnatelli and Van Mons have likewise produced fulminations by concussion, as well with nitrate of mercury and phosphorus, as with phosphorus and most other nitrates. Cinnabar likewise is amongst the substances which, according to MM. Fourcroy and Vauquelin, detonate by concussion with oxymuriate of potash [potassium chlorate].
>
> But mercury and most if not all its oxides, may, by treatment with nitric acid and alcohol, be converted into a whitish crystallized powder possessing all the inflammable properties of gunpowder, as well as many peculiar to itself.

Mercury fulminate

He described the procedure for making it—putting red oxide of mercury into alcohol and then adding nitric acid. The oxide gradually dissolved, and then the mixture boiled, producing dense white smoke and a dark precipitate that eventually turned white. He filtered the white crystals and dried them. In testing these crystals, he poured sulfuric acid on them and was quite surprised at the resulting explosion:

I therefore, for obvious reasons, poured sulphuric acid upon the dried crystalline mass, when a violent effervescence ensued, and, to my great astonishment, an explosion took place.

He then performed more tests:

I first attempted to make the mercurial powder fulminate by concussion; and for that purpose laid about a grain of it upon a cold anvil, and struck it with a hammer, likewise cold: it detonated slightly, not being, as I suppose, struck with a flat blow; for, upon using 3 or 4 grains, a very stunning disagreeable noise was produced, and the faces both of the hammer and the anvil were much indented.

Half a grain or a grain, if quite dry, is as much as ought to be used on such an occasion.

The shock of an electrical battery, sent through 5 or 6 grains of the powder, produces a very similar effect: it seems indeed, that a strong electrical shock, generally acts on fulminating substances like the blow of a hammer. Messrs, Fourcroy and Vauquelin found this to be the case with all their mixtures of oxymuriate of potash.

To ascertain at what temperature the mercurial powder explodes, 2 or 3 grains of it were floated on oil, in a capsule of leaf tin; the bulb of a Fahrenheit's thermometer was made just to touch the surface of the oil, which was then gradually heated till the powder exploded, as the mercury of the thermometer reached the 368th degree.

The next tests were made to test the new crystals in the same way gunpowder was normally tested:

Desirous of comparing the strength of the mercurial compound with that of gunpowder, I made the following experiment, in the presence of my friend Mr. Abernethy.

Finding that the powder could be fired by flint and steel, without a disagreeable noise, a common gunpowder proof, capable of containing eleven grains of fine gunpowder, was filled with it, and fired in the usual way: the report was sharp, but not loud. The person who held the instrument in his hand felt no recoil; but the explosion laid open the upper part of the barrel, nearly from the touch-hole to the muzzle, and struck off the hand of the register, the surface of which was evenly indented, to the depth of 0,1 of an inch, as if it had received the impression of a punch.

The instrument used in this experiment being familiarly known, it is therefore scarcely necessary to describe it; suffice it to say, that it was of brass, mounted with a spring register, the moveable hand of which closed up the muzzle, to receive and graduate the violence of the explosion. The barrel was half an inch in caliber, and nearly half an inch thick, except where a spring of the lock impaired half its thickness.

At this point, it was time to test the new explosive in a real gun:

A gun belonging to Mr. Keir, an ingenious artist of Camdentown, was next charged with 17 grains of the mercurial powder, and a leaden bullet. A block of wood was placed at about eight yards from the muzzle, to receive the ball, and the gun was fired by a fuse. No recoil seemed to have taken place; as the barrel was not moved from its position, although it was in no ways confined. The report was feeble: the bullet, Mr. Keir conceived, from the impression made upon the wood, had been projected with about half the force it would have been by an ordinary charge, or 68 grains, of the best gunpowder. We therefore recharged the gun with 34 grains of the mercurial powder; and, as the great strength of the piece removed any apprehension of danger, Mr. Keir fired it from his shoulder, aiming at the same block of wood. The report was like the first in section IV, sharp, but not louder than might have been expected from a charge of gunpowder. Fortunately, Mr. Keir was not hurt, but the gun was burst in an extraordinary manner.

The breech was what is called a patent one, of the best forged iron, consisting of a chamber 0,4 of an inch thick all round, and 0,4 of an inch in caliber; it was torn open and flawed in many directions, and the gold touch-hole driven out. The barrel, into which the breech was screwed, was 0,5 of an inch thick; it was split by a single crack three inches long, but this did not appear to me to be the immediate effect of the explosion.

I think the screw of the breech, being suddenly enlarged, acted as a wedge upon the barrel. The ball missed the block of wood, and struck against a wall, which had already been the receptacle of so many bullets, that we could not satisfy ourselves about the impression made by this last.

As it was pretty plain that no gun could confine a quantity of the mercurial powder sufficient to project a bullet, with a greater force than an ordinary charge of gunpowder I determined to try its comparative strength in another way.

I procured two blocks of wood, very nearly of the same size and strength, and bored them with the same instrument to the same depth. The one was charged with half an ounce of the best Dartford gunpowder, and the other with half an ounce of the mercurial powder; both were alike buried in sand, and fired by a train communicating with the powders by a small touch-hole.

The block containing the gunpowder was simply split into three pieces: that charged with the mercurial powder was burst in every direction, and the parts immediately contiguous to the powder were absolutely pounded, yet the whole hung together, whereas the block split by the gunpowder had its parts fairly separated. The sand surrounding the gunpowder was undoubtedly most disturbed: in short, the mercurial powder appeared to have acted with the greatest energy, but only within certain limits.

The reason the gun was damaged is that mercury fulminate is a high explosive. Unlike gunpowder, it detonates very rapidly, producing enormous pressures without producing the large quantities of hot gas that gunpowder needs in order to produce the same explosion. This is the reason high explosives do not make good propellants for bullets, cannonballs, or shells—they produce a shock wave that shatters whatever it is in contact with, but does not have the energy to actually move the material very far. A small amount of energy, released in a very small amount of time, has a lot of power. Power is what breaks things, while energy is what moves things. Howard was well aware of this:

> From the experiments related in the 4th and 5th sections, in which the gunpowder proof and the gun were burst, it might be inferred, that the astonishing force of the mercurial powder is to be attributed to the rapidity of its combustion; and, a train of several inches in length being consumed in a single flash, it is evident that its combustion must be rapid. From the experiments of the 6th and 7th sections, it is sufficiently plain that this force is restrained to a narrow limit; both because the block of wood charged with the mercurial powder was more shattered than that charged with the gunpowder, whilst the sand surrounding it was least disturbed; and likewise because the glass globe withstood the explosion of 10 grains of the powder fixed in its centre: a charge I have twice found sufficient to destroy old pistol barrels, which were not injured by being fired when full of the best gunpowder.

Howard went on to try other metals. When he tried silver, he also got an explosive, which he was careful to note is different from the fulminating silver produced by Berthollet (and indeed, they are different compounds). Silver fulminate is even more sensitive than mercury fulminate. It cannot be stored in much quantity, since it detonates under its own weight.

The next experiments were done with the assistance of the

Silver fulminate

local military, where several tests were done by exploding small quantities inside cannons and noting the shattering effects. In one test, a small cannon was destroyed, and in another, a bursting shell was demonstrated:

> Finding that the carronade, from the great comparative size of its bore to that of its length, required a larger quantity of mercurial powder to burst it than we were provided with, we charged a half-

pounder swivel with an ounce and an half avoirdupois of the mercurial powder, (the service charge of gunpowder being 3 ounces) and a half-pound shot between two wads. The piece was destroyed from the trunnions to the breech, and its fragments thrown thirty or forty yards. The ball penetrated five inches into a block of wood, standing at about a yard from the muzzle of the gun; the part of the swivel not broken, was scarce, if at all, moved from its original position.

One ounce avoirdupois of the mercurial powder, enclosed in paper, was placed in the centre of a shell 4,4 inches in diameter, and the vacant space filled with dry sand.

The shell burst by the explosion of the powder, and the fragments were thrown to a considerable distance. The charge of gunpowder employed to burst shells of this diameter, is 5 ounces avoirdupois.

Howard's experiments with mercury fulminate led to his being awarded the Copley Medal of the Royal Society, an award his benefactor, Irish chemist Peter Woulfe, had earned a few years earlier. The experiments also gained him fame both at home in England and across Europe. He went on to study meteorites and showed that they were not of this earth but had fallen from outer space.

As it turns out, Howard was not the first to make mercury fulminate.

Cornelis Jacobszoon Drebbel was a Dutch engraver who later became famous for his many inventions. He made a clock that wound itself using changes in temperature and atmospheric pressure. He modified that mechanism to control the heat of a furnace, thus inventing the thermostat. He applied the same mechanism to an incubator for raising chickens. An accident led to the discovery of a tin mordant for a bright scarlet dye (he had dissolved some tin in aqua regia, and it spilled into some cochineal dye he had planned to use in a thermometer). He made compound microscopes and may have been the first to use two convex lenses in a microscope. He built a submarine rowed by oars and built of leather covering an open-bottomed wood frame.

As an alchemist, he introduced England to a method of making sulfuric acid by burning sulfur with saltpeter. He wrote a paper on the transmutation of elements in 1608 and another alchemical paper in 1621 that discusses extracts of plants, minerals, and metals. In his work for the British Navy, he devised torpedoes and sea mines, using a detonator that has caused some confusion on the part of his biographers. In the same sentence, he is said to use "Batavian tears," *aurum fulminarum*, and fulminating mercury to detonate the mines. Batavian tears are also known as Prince Rupert's Drops. They are teardrop-shaped bits of molten glass cooled quickly in water. When the glass "tail" is broken, the entire drop shatters explosively, due to the high compression the cooling gave to the glass. *Aurum fulminarum* is fulminating gold, which the biographers confuse with mercury fulminate.

While Drebbel's claim to have created mercury fulminate is suspect, the German chemist Baron Johann von Löwenstern-Kunckel described making it in 1690. In his book *Laboratorium Chymicum*, he told of the vigorous reaction that mercury nitrate has with "*spiritus vini*" (ethanol). He described the explosion that resulted but was not known to have used the substance for anything. He also wrote about fulminating gold and fulminating silver, both of which were well known to alchemists of the time. His experiments with gold produced a brilliant red glass, called "ruby-glass," by incorporating gold nanoparticles produced by precipitation from solution. The son of a glassmaker, with a long family history of glassmaking, he built several glass factories under the patronage of Frederick William, Elector of Brandenburg.

By 1807 fulminate of mercury was well known, due to the prestige of the journal where Edward Howard had published. In that year a Scottish minister named Alexander John Forsyth patented a new way to fire a gun, using the new compound. He was an ardent duck hunter, and one of his main complaints was that the noise of the flintlock hammer striking the spark, and the subsequent flash and smoke from the primer pan, frightened the ducks into flight before the powder charge in the gun had time to fire the bird shot.

His invention involved putting a mixture of two fulminating powders, mercury fulminate and the potassium chlorate gunpowder, next to the touchhole of the gun, and modifying the flintlock mechanism to act like a hammer, striking the fulminating powder and igniting the main charge in the gun. The fulminating powder acted very quickly on the gunpowder charge, and the ducks no longer had any warning.

Not much happened with the invention until after the patent ran out. Forsyth turned down an offer by Napoleon Bonaparte of 20,000 pounds sterling to bring his invention to France, and in general seemed to lack the ability to do much in the way of marketing.

After the patent expired, however, many improvements came in rapid succession by a number of different inventors in France, Switzerland, and Britain. The tube-lock held the fulminating powder in a metal tube that was crushed by the falling hammer. It was quickly succeeded by the percussion cap, a metal cup that fit over a nipple at the gun's touchhole, and was likewise crushed by the falling hammer.

In Prussia the Dreyse needle gun used a long needle to pierce a paper cartridge to hit the enclosed percussion cap.

By the 1850s the idea of putting the percussion cap, the powder charge, and the bullet into a single metallic cartridge made loading and firing a gun much faster and easier. The soft brass casing made breech-loading weapons finally practical, as it would expand in the chamber and seal in the exploding gases.

Percussion caps

Black powder percussion cap rifle

The fulminating mixture still had some problems, however. The chlorate in it, and to some extent the mercury fulminate, was corrosive, and the steel barrels of the guns were prone to rust and jam. Later primers used just the mercury fulminate, until the discovery of even better primary explosives, the name given to sensitive compounds that explode on concussion and are used to set off less sensitive, secondary explosives.

In 1858 German chemist Peter Griess discovered diazonitrophenol, a powerful and sensitive explosive, while working with organic compounds rich in nitrogen. He is also thought to be the discoverer (in 1874) of lead styphnate, a now widely used primary explosive for primers in ammunition. The method currently used to produce lead styphnate is due to the work in 1919 by Edmund von Herz. Because it is noncorrosive and less sensitive to shock than mercury fulminate, lead styphnate replaced that compound in ammunition primers.

In 1891 the German chemist Julius Wilhelm Theodor Curtius discovered lead azide. While it reacts with copper, cadmium, and zinc, and alloys of those (such as brass and bronze), it does not corrode steel. It is used as a primer in the same way as lead styphnate and is also used to make bullets that explode on impact. The year before (1890), he discovered another primary explosive, hydrazoic acid (hydrogen azide).

Black powder percussion cap revolver

It is not surprising that the first high explosives were the ones that were very sensitive and would detonate when slightly heated or stirred. An explosion tends to get one's attention. More stable high explosives, such as picric acid, discovered by Irish chemist Peter Woulfe in 1771, and trinitrotoluene (TNT), discovered in 1863 by German chemist Julius Wilbrand, were used as dyes until their explosive nature was discovered much later.

The sensitive primary explosives are a diverse group. There are many ways to make a chemical that comes apart easily, or is made of things that bind much stronger to one another in a different arrangement. An example of a molecule that comes apart easily is *benzvalene*.

Benzvalene

Notice how four of the six carbon atoms are connected by acute bond angles. This causes strain, and the molecule relieves that strain by reconfiguring into simpler smaller molecules explosively.

An example of a molecule whose parts bind tighter when rearranged is lead azide.

Those two groups of three nitrogen atoms could relax into three groups of two nitrogen atoms, with a strong triple bond joining

Lead azide

each pair, thus releasing a lot of energy. Many azides are explosive and usually quite sensitive to heat, friction, and concussion.

Sodium azide is another high explosive that you might find in your garage. It is the explosive used to inflate airbags in cars. A mixture of sodium azide with potassium nitrate and silicon dioxide (silica) is detonated inside the bag when the electronics detect a collision. The highly toxic sodium azide is converted to sodium metal and nitrogen gas. The gas expands the bag. The sodium metal reacts with the potassium nitrate, producing potassium oxide and sodium oxide (and a little extra nitrogen gas). The metal oxides combine with the silica to produce silicate glass, a harmless by-product.

Sodium azide

The amount of explosive in airbags is substantial. In a driver's-side front airbag, there are about 50 grams of sodium azide. The passenger-side airbag is bigger (the passenger is farther away) and contains 200 to 250 grams of explosive. Together, they have about twenty shotgun shells' worth of explosive.

Most primary explosives have both features of an explosive—they come apart easily, and they recombine into smaller parts bound more tightly to one another than they were in the original molecule.

Another feature of many explosive compounds is the presence of an oxidizer on the same molecule as a fuel. As seen with gunpowder, getting the fuel and the oxidizer close together made the product burn faster. By using a better oxidizer, Fourcroy's gunpowder made with potassium chlorate instead of potassium nitrate would explode by concussion. But what if the fuel was on the same molecule as the chlorate oxidizer? The result is something like ammonium chlorate.

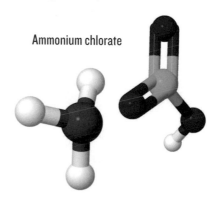

Ammonium chlorate

The chlorine atom has three oxygen atoms to lose, and the nitrogen atom has three hydrogen atoms to lose. When the molecule is heated or hit with a hammer, it produces nitrogen gas, water vapor, hydrogen chloride gas, and a loud bang.

Another example is the class of explosives that combine peroxides with hydrocarbons. A peroxide is two oxygen atoms connected by a weak single bond. That bond is easily broken, and the oxygen atoms are then free to rearrange with the carbon atoms and hydrogen atoms to form carbon dioxide and water vapor. An example is triacetone triperoxide, also known as TATP.

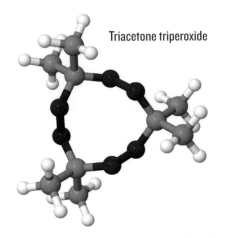

Triacetone triperoxide

Triacetone triperoxide was discovered by German chemist Richard Wolffenstein in 1895. Because it is easy to make from household products, it was used by Palestinian suicide bombers and in several bombings in Europe. It is a sensitive primary explosive and often detonates during manufacture, taking with it the bomb makers.

You can see in the image that there are three acetone molecules connected by three peroxide links. Those links break, and the oxygen atoms combine with the carbon atoms and hydrogen atoms violently.

Another explosive peroxide is one you might have in your medicine cabinet. Benzoyl peroxide is used to bleach flour (peroxides make good bleaching agents) and to clear up acne. In pure form, it is a primary explosive, detonating by heat or shock. In dilute form, it can be applied to the skin, where it breaks down into oxygen, which kills germs by its bleaching action, and benzoic acid, which is a topical antiseptic.

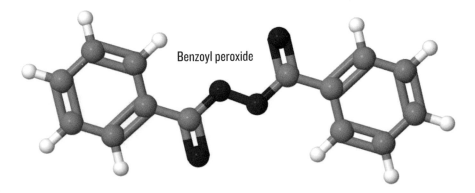

Benzoyl peroxide

Yet another explosive peroxide that you might find in a household cleaning product or in a swimming pool disinfectant is peroxymonosulfuric acid.

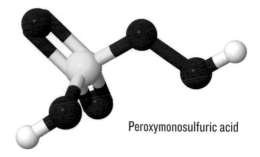

Peroxymonosulfuric acid

Produced by mixing sulfuric acid with hydrogen peroxide, it is one of the strongest oxidizing agents known. An explosive by itself in pure form, it is also capable of adding a peroxide group to compounds of carbon and hydrogen that it touches, making other explosives, such as acetone peroxides. It was first discovered in 1898 by German chemist Heinrich Caro and is sometimes called Caro's acid.

If you work with any of the common two-part plastic resins, you may have a third explosive in your house. Methyl ethyl ketone peroxide (MEKP) is used as a catalyst and hardener for thermosetting polyester plastics. Similar to triacetone triperoxide, the compound is quite explosive and quite sensitive to shock and heat. Fortunately, in dilute form, it does not explode.

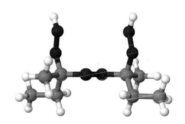

Methyl ethyl ketone peroxide

Because primary explosives are such a diverse group, they include several compounds that are unique or unusual, even for explosives.

One such unique compound is copper acetylide, one of the few explosives that produces no gases when it detonates. It is made by passing acetylene gas through a solution of copper chloride and ammonia. The red precipitate is quite sensitive to heat and shock. Acetylene plants no longer use copper pipes because of the danger of producing this explosive accidentally.

Silver acetylide is also explosive and produces no gas. Silver acetylide mixed with silver nitrate (which adds oxygen to combine with the carbon) is used in some commercial

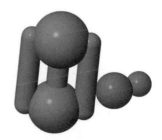

Copper acetylide

explosives. It is a little less explosive than the pure silver acetylide and does produce gases on detonation.

Another interesting special case molecule is xenon trioxide.

Xenon trioxide

Noble gases like xenon do not combine with other atoms easily. Xenon trioxide easily and explosively breaks down into the gases xenon and oxygen. It is an extremely powerful oxidizing agent, and the dry crystals will explode on contact with cellulose and other organic molecules. It will detonate spontaneously at room temperature. Xenon dioxide is also explosive, and xenon tetroxide explodes if it gets above –35.9°C.

Another curious special case is explosive antimony. Antimony crystals can come in two forms. One is a stable metallic form with shiny lustrous crystals. This is called β-antimony. Another is a yellow form, called α-antimony, which changes into the metallic form when heated. If exposed to light, α-antimony turns into a third form, which is black and noncrystalline. This black form turns into the metallic β-antimony when heated.

In 1858 the English electrochemist George Gore (one of the many inventors of the safety matches used today), was electroplating antimony from a solution of antimony trichloride onto a piece of copper. This produced a fourth form of antimony. It is a solid solution of antimony trichloride in α-antimony. It has the appearance of shiny gray graphite.

When he scraped some of it off the copper electrode for analysis, it exploded.

The α-antimony releases a good deal of heat when it transforms back into the shiny β-antimony form. This raises the temperature to 250°C, and vaporizes the antimony trichloride in a puff of toxic white smoke. When crushed in a mortar and pestle, it detonates with a loud crack.

It is perhaps not surprising that the invention of an exploding compound might change history. The subtlety by which it happens, however, is interesting. In 1812 the French chemist Pierre Louis Dulong was experimenting with chlorine gas and ammonium nitrate and purifying the reaction products. One of those reaction products is nitrogen trichloride. It is a very sensitive high explosive, and in two explosions, he lost two fingers and an eye (one might ask why he wasn't much more careful after the first explosion, but that bit of history doesn't seem to have survived the passage of time).

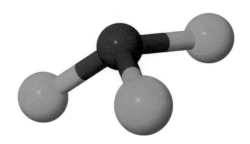

Nitrogen trichloride

A year later, the Cornish chemist and newly knighted Sir Humphry Davy, former lecturer at the Royal Institution, also damaged his eyesight in a nitrogen trichloride explosion. As the discoverer of the elements potassium, sodium, calcium, magnesium, boron, and barium, he was the person who gave the name chlorine to the gas Swedish chemist Carl Wilhelm Scheele had discovered in 1774, and insisted it was an element.

Davy describes the new explosive in the *Philosophical Transactions* of the Royal Society on November 5, 1812:

I immediately exposed a phial, containing about six cubical inches of chlorine, to a saturated solution of nitrate of ammonia, at the temperature of about 50° in common day-light. A diminution of the gas speedily took place; in a few minutes a film, which had the appearance of oil, was seen on the surface of the fluid; by shaking the phial is collected in small globules, and fell to the bottom. I took out one of the globules, and exposed it in contact with water to a gentle heat: long before the water began to boil, it exploded with a very brilliant light, but without any violence of sound. . . .

An attempt was made to procure the substance in large quantities, by passing chlorine into Wolfe's bottles, containing the different solutions, but a single trial proved the danger of this mode of operating; the compound had scarcely began to form, when, by the action of some ammoniacal vapour on chlorine, heat was produced, which occasioned a violent explosion, and the whole apparatus was destroyed.

I attempted to collect the products of the explosion of the new substance, by applying the heat of a spirit lamp to a globule of it, confined in a curved glass tube over water: a little gas was at first extricated, but long before the water had attained the temperature of ebullition, a violent flash of light was perceived, with a sharp report; the tube and the glass were broken into small fragments, and I received a severe wound in the transparent cornea of the eye, and obliges me to make this communication by an amanuensis. This experiment proves what extreme caution is necessary in operating on this substance, for the quantity I used was scarcely as large as a grain of mustard seed. . . .

The mechanical force of this compound in detonation, seems to be superior to that of any other known, not even excepting the ammoniacal fulminating silver. The velocity of its action appears to be likewise greater.

The damage to his eyes caused him to hire a valet, who could also assist him in his experiments. That young man was Michael Faraday, who went on to become an even more famous scientist (thus the subtle effects of explosions on history). And, in proof that even brilliant people don't learn from their mistakes, both Faraday and Davy were injured in yet another experiment with nitrogen trichloride, although this time some precautions were taken:

The action of mercury on the compound appeared to offer a more correct and less dangerous mode of attempting its analysis; but on introducing two grains under a glass tube filled with mercury and inverted, a violent detonation occurred, by which I was slightly wounded in the head and hands, and should have been severely wounded, had not my eyes and face been defended by a plate of glass attached to a proper cap, a precaution very necessary in all investigations of this body.

Soon after Humphry Davy hired Faraday, the two went on a trip to France to receive a medal from Napoleon Bonaparte for Davy's work in electrochemistry. There, the famous French chemist Joseph Louis Gay-Lussac told him about a new substance discovered by a fellow French chemist named Bernard Courtois.

Courtois owned a factory that produced potassium nitrate for Napoleon's war efforts. The gunpowder factories were having trouble finding enough wood ash with which to make the nitrate. Courtois turned to using seaweed ashes, since seaweed was abundant on the French coast. Courtois was trying to find out what was causing the corrosion on the copper pots used to process the seaweed ash. He added sulfuric acid to the ash and noticed a peculiar purple vapor.

Davy writes about the "peculiar substance" in the *Philosophical Transactions* of the Royal Society on January 20, 1814:

> A new and very curious substance has recently occupied the attention of chemists at Paris.
>
> This substance was accidentally discovered about two years ago by M. Courtois, a manufacturer of saltpeter in Paris. In his processes for procuring soda from the ashes of sea weeds (*rendres de vareck*) he found the metallic vessels much corroded; and in searching for the cause of this effect, he made the discovery. The substance is procured from the ashes, after the extraction of the carbonate of soda, with great facility, and merely by the action of sulfuric acid: —when the acid is concentrated, so as to produce much heat, the substance appears as a vapour of a beautiful violet color, which condenses in crystals having the colour and the luster of plumbago.
>
> M. Courtois soon after he had discovered it, gave specimens of it to M. M. Desormes and Clement for chemical examination; and those gentlemen read a short memoir upon it, at a meeting of the Imperial Institute of France, on Nov. 29th. In this memoir, these able chemists have described its principal properties; they mentioned that its specific gravity was about four times that of water, that it becomes a violet coloured gas at a temperature below that of boiling water, that it combines with the metals and with phosphorus and sulphur, and likewise with the alkalies and metallic oxides, that it forms a detonating compound with ammonia.

Davy performs his own experiments and concludes that the "curious substance" is not a compound but an "undecompounded body" (i.e., an element), which he names iodine.

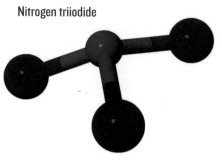

Nitrogen triiodide

The detonating compound is nitrogen triiodide.

Nitrogen triiodide is the only known explosive that is so sensitive it will detonate when hit by alpha particles (helium nuclei produced by radioactive elements). Chemistry professors love to demonstrate how it can be set off using the touch of a feather.

GUNCOTTON AND
SMOKELESS POWDERS

The explosive compounds that are very sensitive were discovered first, since once you have made something that blows up in your face, you tend to notice that special property. For less sensitive explosives, it can take some time between the discovery of the substance and the identification of its explosive potential. This is the case for many explosives and will be evident in explosives discussions to follow. It is certainly the case with guncotton (nitrocellulose).

Sometime around the year 1310, an alchemist named Paul of Taranto published a series of books on his trade, under the name Geber, seeking to gain stature for the work by attributing it to the famous Arabian alchemist of that name. In those books, he describes the production and use of several new compounds, including nitric acid, aqua regia, sulfuric acid, and silver nitrate.

Nitric acid, especially when combined with concentrated sulfuric acid, came to be a very effective way of adding nitrate and nitro groups to organic compounds to produce high explosives.

Henri Braconnot was a French chemist and inventor. He had learned chemistry in part by attending the lectures of comte de Fourcroy just after the turn of the nineteenth century, during a brief (two year) stay in Paris. Specializing in the chemistry of plants, he is known mainly for the discovery of the plant structural molecules pectin and chitin, the sugar glucose, the amino acid glycine, and several organic acids. He comes to this discussion for some experiments he performed by treating plant material with acids in 1832.

The discovery of glucose came through the addition of wood, straw, and cotton to sulfuric acid. When Braconnot tried nitric acid on starch, he found he had produced a flammable substance he could dissolve in acetic acid to produce a glassy varnish. He named the substance *xyloïdine*. He found that sawdust, cotton, and linen produced the same substance when nitrated the same way. Today the substance is known as nitrocellulose, and the varnish (nitrocellulose dissolved in ether) is called celluloid.

Nothing much came of his discovery.

Théophile-Jules Pelouze was another French chemist, who, despite being the son of an industrial chemist and a laboratory assistant to Gay-Lussac, is not known for any great discoveries or inventions. But he did play with adding nitric acid to cardboard and paper, making nitrocellulose in 1838. He is mostly known today for his two students, Ascanio Sobrero, who discovered nitroglycerin, and Alfred Nobel, who tamed nitroglycerin with his invention of dynamite.

Pelouze experimented with Braconnot's xyloïdine and noted that it ignited at 180°C, burned violently, and left no residue. He experimented with nitrating paper and produced a waterproof parchment-like material that he thought was paper coated with xyloïdine but was in reality nitrocellulose all the way through. However, although he did his experiments believing the material would be useful in artillery, he did not commercialize his results.

A third French chemist, Jean-Baptiste Dumas, also played with adding nitric acid to cellulose, calling his creation *nitramidine*. But he also did nothing with it.

The main reason these three chemists ignored nitrocellulose as a propellant or an explosive is that their mixtures were highly unstable. The acid left in the fibers would continue to act on them, generating heat and sometimes premature detonation or the material would break down and lose its effectiveness. Not to mention what all that nitric acid would do to a gun barrel.

Then, in 1846, the story goes, a Swiss-German chemist named Christian Friedrich Schönbein was working at home and spilled some concentrated nitric acid on the kitchen table. Grabbing his wife's cotton apron, he mopped up the spill. As he dried the apron on the door of the stove, it flashed briefly and disappeared.

The story may be apocryphal. Schönbein had discovered ozone and published his discovery in 1840, and in 1847 published a paper, "On the Discovery of Gun-Gotten," describing his work with nitrating cane sugar and implying that his work was prompted by his earlier work with ozone. To nitrate the sugar, he mixed two parts of sulfuric acid with one part of nitric acid, and stirred in finely powdered sugar in an ice bath at 0°C. The thick mass that resulted was then washed thoroughly to remove the acid and then slowly dried without overheating.

The resulting nitro-sucrose was brittle until warmed but then became plastic and finally runny at the boiling point of water. When heated too much, it disappeared with a flash. He then experimented with other nitrated organic substances.

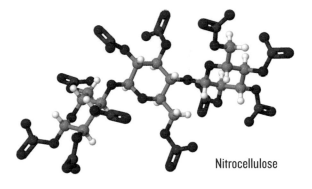

Nitrocellulose

The substance to which I have given in German the name of schiess-wolle, and in English that of gun-cotton, having excited a lively curiosity, it may be interesting to the scientific world to become acquainted with some details of the way in which I was first led to its discovery.

The results of my researches on ozone led me in the course of the last two years to turn my attention particularly to the oxyds of nitrogen, and principally to nitric acid.

Schönbein goes on to describe his theory that hydrated sulfuric acid was a mixture of SO_2 and HO_2, and not SO_3 and HO as the prevailing theory held. He then goes on describing the HO_2 part:

My experiences with ozone having shown that this body, which I consider to be a distinct peroxyd of hydrogen, forms, as well as chlorine, at the ordinary temperature, a peculiar compound with olefiant gas, without apparently oxydizing in the least either the hydrogen or the carbon of this gas, I had the idea that it would not be impossible that certain organic matters, exposed to a low temperature, would likewise form compounds, either with the peroxyd of hydrogen alone, which, on my hypothesis, occurs in a state of combination or of mixture in the acid mixture, or with NO_4.

At this point he describes his work nitrating sugar and then other organic substances.

In March, I sent specimens of my new compounds to some of my friends, in particular to Messrs. [Michael] Faraday, [Sir John] Herschel, and [Sir William Robert] Grove. It is necessary to note expressly that the gun-cotton formed part of these products; but I must add, that hardly was it discovered when I employed it in experiments of shooting, the success of which encouraged me to continue them.

It must have been nitrocellulose's time, since two more German chemists also made the compound in the same year.

Schönbein describes being the first to fire a cannon using the new guncotton, after testing it in smaller mortars to ensure it could safely be done. He then used it to blast some rocks and blow up some old walls, proving to himself that his invention was superior to gunpowder for those purposes. He went to England and experimented with it in Cornish mines and with many small arms. In his paper, he is careful to distinguish his guncotton from Braconnot's xyloïdine, saying the latter had too much carbon and not enough oxygen to be a good explosive.

Washing the acid from the product was not immediately recognized as being important. In the same year as the latest discovery (1846), a plant was built in England to manufacture guncotton using English patents based on Schönbein's discovery. Due to a lack of safety precautions, the plant exploded that July, killing twenty-one of the workers. The company Schönbein had contracted with to produce the explosive, John Hall & Sons, refused to continue manufacturing it. Two other guncotton plants in France also exploded, and no more attempts to produce guncotton in either England or France were made for another fifteen years.

Schönbein then contacted the Deutscher Bund, the loose confederation of thirty-nine German states in central Europe that replaced the former Holy Roman Empire. A committee studied his proposal for years, finally deciding to take no action. On the advice of Austria's representative to the committee, Baron Wilhelm Freiherr Lenk von Wolfsberg, a major general and owner of the Corps Artillery Regiment, Austria then purchased the rights to the process for a price much below what Schönbein had originally asked.

In 1853 von Lenk (as the baron is usually referred to) developed a much more elaborate method for washing and preparing guncotton, involving washing it for three weeks, then boiling it in alkali, mixing it with sodium silicate, and then drying it. He tried his new product in iron guns but the mixture damaged them too much. In bronze guns, which didn't shatter as easily, he had better success. He experimented with exploding shells filled with guncotton but these often burst inside the cannon, due to the sudden acceleration provided by the new propellant. He had better luck sending the guncotton shells toward the target when he used old-fashioned black powder in the cannons.

The Austrian experiments ended in 1863, when the manufacturing plant exploded in Hirtenberg, and the Austrian government abandoned guncotton manufacture.

It took years before the British chemist Sir Frederick Augustus Abel came up with a safer way to manufacture the explosive and production resumed in 1865. Abel had studied chemistry at the Royal Polytechnic Institution and the Royal College of Chemistry, and in 1852 at age twenty-five he was appointed lecturer in chemistry at the Royal Military Academy, a post previously occupied by Michael Faraday.

Abel used a pulping process to aid in the rinsing and produced the nitrocellulose in a more useful form. Several other researchers had been experimenting with guncotton and had established that thorough rinsing was needed, and sometimes neutralization with alkalis. Carding the resulting fibers during the rinse was also recommended. This proved to be key, as separating the fibers allowed a better rinse. In the intervening fifteen years, explosions at plants in Vienna and elsewhere caused further moratoria on its manufacture.

Abel's 1865 patent notes the method of pulping:

The fibre is then taken in the wet state and converted into pulp in the same manner as is practiced by paper-makers, by putting the fibre into a cylinder, having knives revolving rapidly, working close to fixed knives.

In Abel's patent, the guncotton is not used by itself as a propellant. Instead, it is added to a form of gunpowder comprising 65% sodium nitrate, 16% charcoal, 16% sulfur, and 3% guncotton.

Abel published a paper titled "Researches on Gun-Cotton—on the Manufacture and Composition of Gun-Cotton" in the *Philosophical Transactions* of the Royal Society on April 10, 1866, detailing his study of the new propellant. In that paper, he describes in detail the chemistry of the compound, but also how he came to study it:

Early in 1863, by desire of the Secretary of State for War, I entered upon a detailed investigation of the manufacture of gun-cotton, the composition of the material when produced upon an extensive scale, its behaviour under circumstances favourable to its change, and other subjects relating to the chemical history of this remarkable body.

Abel credits von Lenk with his "persevering endeavors to perfect this material," but goes on to say that von Lenk's process "does not, at first sight, present any important features of novelty." Then he continues:

But the application, upon a manufacturing scale, of the system which has been pursued under Lenk's direction at Hirtenberg, brings to light several details of considerable value, the adoption of which unquestionably ensures the attainment of greater uniformity and purity of the product of manufacture than could be secured by the earlier modes of operation.

Abel takes issue with von Lenk's three-week washing period, not finding any improvement in the product over a forty-eight-hour washing. But the main improvement is the pulping, because cotton fibers are hollow and tubular, and the acid is difficult to remove from inside the tubes. Pulping roughens the ends of the fibers, opening the tubes to the rinse water and to the later alkali bath.

I have been using the terms *nitrocellulose*, *nitroglycerin*, and *nitrosucrose* in this book because that is what they are called in industry, and that is how the public knows them. A chemist, however, would make the distinction between a true nitro compound (like trinitrotoluene, or TNT) and these compounds, which are more properly called nitrates. Thus, the proper names are cellulose trinitrate, glyceryl trinitrate, and sucrose trinitrate.

The distinction has to do with one of the oxygen atoms. The image above is nitric acid. On the right side of the molecule, there is a red oxygen atom between the blue nitrogen atom and the white hydrogen atom. That oxygen atom is the difference between a nitrate group (the part of the nitric acid without the hydrogen) and a nitro group (the part of the nitric acid without the oxygen and the hydrogen on the right).

Nitric acid

Pure guncotton was still too *brisant* (shattering) to use in guns or cannon by itself. It was, however, well suited to mining, where the rock is *supposed* to shatter. Abel's method produced a stable form due to the removal of the acid. It did not deteriorate, and it was not affected by moisture, unlike black powder.

Another use for guncotton was in naval mines. A primary explosive, such as mercury fulminate, could detonate a small amount of dry guncotton, which could then detonate a larger charge of wet guncotton. This use of a "booster" charge would be used later with other explosives.

To make guncotton suitable as a propellant in firearms, its combustion had to be slowed. The first practical smokeless powder was invented in 1864 by a captain in the Prussian artillery named Johann F. E. Schultze. It was a mix of guncotton made from wood, mixed with potassium nitrate or barium nitrate. The wood contained other materials than cellulose, which, along with the nitrates, slowed the burning of the powder. Still, it was too powerful for rifles (where the bullet resists leaving the gun due to the rifling, so too much pressure builds up) but was suitable for shotguns, due to their smooth bores. Later varieties included various other adulterants to control the rate of burn.

In 1884 the French physicist Paul Vieille invented a smokeless powder that could be used in rifles. He controlled the burn rate by dissolving some of the nitrocellulose in a mixture of ether and alcohol. The remaining undissolved fibers mixed with the dissolved compound and formed a pasty jelly that could be rolled into sheets and cut, or extruded and cut like noodles.

The reason black powder produces so much smoke is that not all of the reaction products are gases. The potassium sulfides, sulfates, and oxides, along with unburned carbon from the charcoal, form a fine powder that disperses in air and collects on the inside of the gun barrels, causing fouling. Nitrocellulose has no metals in it—what smoke it emits is primarily from unburned carbon. The reaction products are primarily nitrogen, water vapor, and oxides of carbon.

Smokeless powder had advantages other than being smokeless. Less of it was needed since it was more powerful than black powder. To get the same effect in a gun, only about a third as much smokeless powder is needed as black powder. Moreover, smokeless powder was far less affected by moisture.

Further improvements to smokeless powder involved coating the extruded pellets with graphite. This slowed the burn rate a bit but just as importantly it made the powder electrically conductive, so static electricity would not build up and create sparks during transport and handling. The extruded material allowed manufacturers to control the surface area to volume ratio, which gave them precise control over the burn rate.

While guncotton is far less sensitive to moisture than black powder is, it is still important to control how it reacts to humidity. Consistent performance as a propellant is necessary if the projectiles are going to hit their mark reliably, and how much moisture is in the powder has an effect on its performance. Guncotton was treated with hot water to create cracks in the outer skin of the colloidal extrusions, where water could collect and remain. If the grains were too dry, they would absorb moisture from the air. If they were too wet, they would dry out in transport. With the minute cracks in the surface, there was just the right amount of water, so the powder neither lost nor gained moisture during storage and transport.

Controlling the rate of burning is very important. The different extruded shapes of the grains are designed to have specific effects on the rate of burn. Simple cylinders, like that of cordite (named because it was extruded in strings or cords), have surfaces that get smaller as the grains burn from the outside in. This means that the rate of burning decreases as the charge burns.

If extruded into strips, flakes, or tubes, the powder burns at a more consistent rate. To get the powder to burn faster as the burn progresses, the grains can be extruded with multiple hollow tubes inside. In a gun, as the bullet travels down the barrel and the volume behind it increases, a steadily increasing pressure caused by faster and faster burning keeps the pressure constant. This leads to faster velocities for the projectile.

The flash of a gun at night can give away its position. A 12-inch gun can project a muzzle flash as far as 150 feet, and this can be seen reflecting off low clouds for 30 miles, farther than the sound of the gun carries.

The flash is caused by superheated air, and by hot uncombusted gases igniting when they encounter oxygen in the air. To reduce the flash, so-called flashless powders were developed. (The use of the term "powder" has continued despite the extruded macaroni-like form of the new propellants.)

To reduce the temperature of the escaping gas, explosives that burn cooler can be added. Ammonium nitrate is one example, as is guanidinium nitrate and nitroguanidine. To prevent the gases from igniting, salts such as potassium chloride can be added that prevent the gases from combining with oxygen while they are hot enough to ignite. Adding inert materials to the mixture that absorb heat and produce gas, such as carbonates that produce carbon dioxide when heated, can cool the output at the same time as they increase the amount of gas produced.

NITROGLYCERIN

In 1846, the same year that Schönbein was creating nitrocellulose, the Italian chemist Ascanio Sobrero was also working with the mixture of sulfuric and nitric acids known to add nitro groups (NO_2) to organic molecules like cellulose, sugar, starch, and others.

The son of a doctor, Sobrero studied medicine in Torino (Turin), and had been authorized to practice as a physician in 1834, when he was twenty-two years old. Following his father's example of a physician who taught at the university, Sobrero took the examination to be a teacher, but due to political differences he was classified as unsuitable.

Disappointed, he decided to leave medicine for the study of chemistry. He worked for about four years as an assistant in an Italian laboratory of general chemistry and then a laboratory of pharmaceutical chemistry. His uncle, Baron General Carlo Rafaele Sobrero, was the general director of the Chemical Laboratory of the Arsenal in Torino, and had studied chemistry at Paris at the *École Polytechnique*. His uncle had become acquainted with several famous European chemists, such as Théophile-Jules Pelouze and Jöns Jacob Berzelius.

With a letter of recommendation from his uncle, Sobrero left for France, where he studied under Théophile-Jules Pelouze and became his assistant in the Frenchman's private laboratory from 1840 to 1843. Pelouze had experimented with nitrating cellulose in 1838 and was still actively investigating the action of nitric acid on organic substances during the time Sobrero was assisting him.

Sobrero left Paris in 1843 to study under the famous German chemist Justus Freiherr von Liebig at the University of Giessen. Liebig has been called one of the greatest chemistry teachers of all time. He was instrumental in developing the study of organic chemistry and the modern laboratory-based teaching of chemistry. For a chemistry student interested in teaching, this was a perfect opportunity.

Having been away from Italy for three years, Sobrero returned to Torino, where he worked again as a general chemist. In 1845 he received the appointment to the chair of chemistry at a new school.

Sobrero had been working with nitric acid and glycerin to produce oxalic acid, and had added nitric acid to the oil from birch trees, following work he had done with Pelouze. When Schönbein published his results with guncotton, Sobrero tried the new recipe (two parts sulfuric acid to one part nitric, a mix where the sulfuric acid acts as a catalyst for the nitration reaction) on glycerin. He also successfully made explosive compounds by nitrating the sugars mannose, sucrose, and lactose.

He reported his findings in a letter to Pelouze in Paris, who read the letter to the *Académie des Sciences* there in 1847, a year after the discovery. In the letter he cautioned about tasting the new chemical, since when he (and others in the lab) tried this on several occasions to test its toxicity, they all got severe headaches. As a vasodilator, nitroglycerin dilates capillaries in the brain, causing the headaches. By the end of the century, nitroglycerin was being used by physicians everywhere to treat the pain of severe angina and is still used by heart patients today.

When Sobrero added the acids to glycerin, he saw a violent reaction take place with lots of red fumes. However, if he carefully added the glycerin to the acid instead, while stirring it in an ice bath at 0°C, the glycerin dissolved. When the acid solution with dissolved glycerin was then added to water, nitroglycerin precipitated out, sinking under the water to the bottom of the dish.

He washed the oily residue in more water and then dried it in a vacuum over sulfuric acid (which has a great affinity for water and thus helps desiccate things). The result was something with the appearance of olive oil (pure nitroglycerin has no color). In a paper

Nitroglycerin

titled "Some New Fulminating Products Obtained by the Action of Nitric Acid on Some Vegetable Organic Substances," which he later presented at the Ninth Italian Scientific Conference in Venice, he describes the new compound:

> It detonates when brought into contact with metallic potassium, and evolves oxides of nitrogen on contact with phosphorus at 20° to 30° C, but at higher temperatures it ignites with an explosion. . . . When heated, nitroglycerin decomposes. A drop heated on platinum foil ignites and burns fiercely. It has, however, the property of detonating under certain circumstances with great violence. On one occasion a small quantity of an ethereal solution of nitroglycerin was allowed to evaporate in a glass dish. The residue of nitroglycerin was certainly not more than 2 or 3 centigrams. On heating the dish over a spirit lamp a most violent explosion resulted, and the dish was broken into atoms. . . .
>
> The safest plan for demonstrating the explosive power of nitroglycerin is to place a drop upon a watch glass and detonate it by touching it with a piece of platinum wire heated to low redness. Nitroglycerin has a sharp, sweet, aromatic taste. It is advisable to take great care in testing this property. A trace of nitroglycerin placed on the tongue, but not swallowed, gives rise to a most violent pulsating headache accompanied by great weakness of the limbs.

Sobrero thought that one of his nitrated sugars, *nitromannite*, might make a good explosive for use in percussion caps. But in 1853 an explosion of 400 grams of the material in a laboratory at his uncle's arsenal caused such damage that he stopped pursuing the project. This may have also led to him not pursuing nitroglycerin as a commercial explosive.

That job ended up with Alfred Bernhard Nobel.

Nobel's father, Immanuel had, after several failed business attempts, finally done well as a manufacturer of machine tools and explosive mines, after moving to St. Petersburg, Russia, from his native Sweden. The now prosperous family could afford private tutors for Alfred and his brothers. By age sixteen Alfred was a knowledgeable chemist, having been tutored by chemist Nikolai Zinin (known as the father of Russian chemistry, who also taught Mendeleev), and fluent in five languages.

In 1850, at the age of seventeen, he spent a year in Paris studying chemistry at the laboratory of Théophile-Jules Pelouze, three years after Pelouze had read Sobrero's paper to the French Academy. He then traveled in Italy, Germany, and the United States, where he worked for John Ericsson, the Swedish engineer who later built the ironclad ship *Monitor* for Union forces in the US Civil War.

When the Crimean War broke out in October 1853, Immanuel Nobel's company in Russia was making a new type of underwater mine, experimenting unsuccessfully with the newly invented nitroglycerin but succeeding with gunpowder. These mines were used in the Baltic to defend Russian ports from the combined French and British fleets. Alfred had returned to St. Petersburg in 1852 and worked with his father during the war. Regarding nitroglycerin, he later recalled:

> The first time I saw nitroglycerine was in the beginning of the Crimean War. Professor Zinin in St. Petersburg exhibited some to my father and me, and struck some on an anvil to show that only the part touched by the hammer exploded without spreading. His opinion was that it might become a useful substance for military purposes, if only a practical means could be devised to explode it. . . . My father tried to explode it during the Crimean war, but completely failed to do so. . . . My father's later experiments with gunpowder mixed with nitroglycerine were all on a small scale.[2]

When the war ended in 1856 with Russia's defeat, the bottom fell out of the Russian exploding mine business, and by 1859 Immanuel Nobel was once again bankrupt and left Russia to return to his native Sweden. Alfred and two of his brothers remained in St. Petersburg, doing mechanical engineering work. Alfred carried on experimenting with nitroglycerin in St. Petersburg. He finally got nitroglycerin to explode underwater by

---

[2] Birgitta Lemmel, "Alfred Nobel—St. Petersburg, 1842–1863," Official Website of the Nobel Prize, www.nobelprize.org/alfred_nobel/biographical/articles/russia/#footnote.

surrounding a glass tube full of the liquid with gunpowder in a zinc can and using a fuse to set off what was basically a firecracker.

In 1863 he left St. Petersburg to rejoin his father in Stockholm, where he continued his experiments with nitroglycerin. He obtained a patent for improvements in the method of producing nitroglycerin as an industrial explosive and for the blasting cap.

In 1864 the small shed he used for manufacturing the blasting oil exploded, killing his younger brother Emil and several others. Undaunted, he continued his research. In 1865 he improved on the blasting cap and moved the company to Krümmel, near Hamburg, in Germany, but the plant in Krümmel also exploded. Taking his experiments onto a raft in the river Elbe, he found that a highly absorbent diatomaceous earth could hold so much nitroglycerin that it will still explode, even when the earth still has the form of a powder. He called his invention dynamite. Later in that year he started the United States Blasting Oil Company in America.

In Nobel's 1866 US patent application for his blasting powder, he not only describes the substance and how to make it but also how to use it and detonate it. He even inserts a plug for his other invention, the blasting cap:

TO ALL WHOM IT MAY CONCERN:

Be it known that I, Alfred Nobel, of the city of Hamburg, Germany, have invented a new and useful Composition of Matter, to wit, an Explosive Powder.

The nature of the invention consists in forming out of two ingredients long known, vis, the explosive substance nitro-glycerine, and an inexplosive porous substance, hereafter specified, a composition which, without losing the great explosive power of nitro-glycerine, is very much altered as to its explosive and other properties, being far more safe and convenient for transportation, storage, and use, than nitro-glycerine.

In general terms, my invention consists in mixing with nitro-glycerine a substance which possesses a very great absorbent capacity, and which, at the same time, is free from any quality which will decompose, destroy, or injure the nitro-glycerine, or its explosiveness.

It is undoubtedly true, as a general rule, that nitro-glycerine, when mixed with another substance, possesses less concentration of power than when used alone; but while the safety of the miner (to prevent leakage into seams in the rock) prohibits the use of nitro-glycerine without cartridges, which latter must of course be somewhat less in diameter than the bore-holes which are to contain them, the powder herein described can be made to form a semi-pasty mass, which yields to the slightest pressure, and thus can be made to full up the bore-hole entirely. Practically, therefore, the miner will have as much nitro-glycerine in the same height of bore-hole with this powder as with nitro-glycerine in its pure state.

This is the real character and purpose of my invention; and in order to enable others skilled in the art to which it appertains (or with which it is most nearly connected) to make, compound, and use the same, I will proceed to describe the same, and also the manner and process of making, compounding, and using it, in full, clear, and exact terms.

The substance which most fully meets the requirements above mentioned, so far as I know or have been able to ascertain from numerous experiments, is a certain kind of silicious earth, or silicic acid, found in various parts of the globe, and known under the several names of silicious marl, tripoli, rotten-stone, &c. The particular variety of this material which is best for my compound is homogeneous, has a low specific gravity, great absorbent capacity, and is generally composed of the remains of infusoria.

So great is the absorbent capacity of this earth, that it will take up about three times its own weight of nitro-glycerine and still retain its powder-form, thus leaving the nitro-glycerine so compact and concentrated as to have very nearly its original explosive power; whereas, if another substance, having less absorbent capacity, is used, a correspondingly less proportion of nitro-glycerine will be absorbed, and the powder be correspondingly weak or wholly inexplosive.

For example, most chalk will take but about fifteen percent of nitro-glycerine and retain its powder-form. Twenty per cent, will reduce it to a paste.

Porous charcoal has also a considerable absorbent capacity, but it has the defect of being a combustible material, and also of less elasticity of its particles, which renders it easy to squeeze out a part of its nitro-glycerine.

The two materials are combined in the following manner:

The earth, thoroughly dried and pulverized, is placed in a wooden vessel. To it is introduced the nitro-glycerine in a steady stream so small that the two ingredients can be kept thoroughly mixed.

The mixing may be effected by naked hand, or by any proper wooden instrument used in the hand, or by wooden machinery.

Sufficient nitro-glycerine should be used to render the compound explosive, but not so much as to change its form of powder to a liquid or pasty consistency.

Practically, about sixty parts, by weight, of nitro-glycerine to forty of earth, forms a useful minimum, and seventy-eight parts, by weight, of nitro-glycerine to twenty-two of earth, the useful maximum of explosive power. The former has a perfectly dry appearance, the latter is pasty.

Between these two extremes the composition will be explosive powder, and will be more easily exploded, and its explosive power greater, as the relative proportion of the nitro-glycerine is greater.

The proportions, by weight, of seventy five of nitro-glycerine and twenty-five of earth, gives a powder as well adapted to ordinary practical purposes as that from any proportions I am now able to name, and can be easily compressed to a specific gravity nearly equal to that of pure nitro-glycerine.

When the mass has been intimately mixed and thoroughly incorporated by stirring and kneading, it is rubbed through a hair, silk, or brass-wire sieve, (iron corrodes) and any lumps which may remain are rubbed with a stiff-bristle brush till they are reduced and made to pass through the sieve.

The powder is then finished and ready for use.

The fineness desired for the powder will determine the fineness of the sieve to be used.

The chief characteristic of this powder is its nearly perfect exemption from liability to accidental or involuntary explosion.

It is far less sensitive than the nitro-glycerine to concussion, and contained in its usual packing, (a wooden cask or box) the latter may be smashed completely to pieces without any danger of an explosion.

Unlike gunpowder, in the open air or in ordinary packing, (a wooden cask or box,) it burns up, when set fire to, without exploding. It can, therefore, be handled, stored, and transported with less danger than ordinary gunpowder.

When confined in a tight and strong enclosure it explodes by heat applied in any form when above the temperature of 360°F. Under all other circumstances it may be exploded by some other explosion in it or into it.

The most simple and certain method known to me of exploding it is as follows:

The end of a common blasting-fuse is inserted into a percussion-cap, and the rim of the cap crimped tightly and firmly about the fuse by nippers, or other means, so as to leave the fulminating-powder of the cap and the end of the fuse tightly and firmly enclosed together. The end of the fuse, with the cap attached, is then embedded in the powder—the more firmly, the more certain the explosion.

In blasting, the powder is pressed tightly about the cap and fuse, and tamping, of sand or other proper material, added, and pressed but not pounded in. A tamping firmly pressed is as good as if rammed in the most solid manner.

The fuse explodes the cap, and this explosion explodes the powder.

I will add here that by carefully packing the end of a good fuse amidst the powder of a charge enclosed, like a blasting charge, in a tight place, the fuse alone will explode the powder, especially if the powder is strongly charged with nitro-glycerine. But this method of explosion requires too much care, and is too uncertain to be depended upon or generally used.

As before stated, the more strongly the powder is charged with nitro-glycerine the more easily it explodes. If, therefore, the powder contains a low proportion of nitro-glycerine, it is necessary to employ in its explosion a correspondingly long, strong, and heavily-charged percussion-cap, made especially for the purpose. For the sake of certainty of explosion it is better to use such a cap in all cases.

If the fire from the fuse comes into contact with the powder before the cap is exploded, which is liable to occur if the fuse is leaky and the cap extends too far into the powder, a portion of the powder will be burned before the explosion takes place. To guard against this, the cap should only be fairly inserted into the powder, and poor fuses wound next to the cap firmly with strong glued paper or hemp, or otherwise secured.

The bore-holes, as a practical but not absolute rule, should be about one-half the size, and the charge should be one-fifth to one-tenth the quantity ordinarily used in gunpowder-blasting.

A very convenient form in which to use the powder is to pack it firmly in cartridges of strong paper.

Having thus described my invention, what I claim is new, and desire to secure by Letters Patent, is—

The composition of matter, made substantially of the ingredients and in the manner and for the purposes set forth.

ALFRED NOBEL

Although the addition of diatomaceous earth to nitroglycerin made the product, dynamite, much safer to use, store, and transport, it was not perfect. It still contained 25% inert material that did not aid in the explosion and took up space in the mining boreholes, thus making it less powerful than nitroglycerin. In addition, after prolonged storage some of the nitroglycerin would wick out of the paper cartridges, where it was once again subject to explode by concussion. Both this "sweating" and the inert ingredients were problems Nobel solved in 1875 with his invention of gelignite.

A mixture of guncotton dissolved in ether and alcohol makes a plastic called collodion. At the time, collodion was used as a wound dressing. Another plastic, celluloid, was made by dissolving guncotton in camphor.

Nobel found that guncotton would dissolve in nitroglycerin. If 7% to 8% guncotton was dissolved in the liquid, the result was a solid jelly that did not sweat, and did not contain any inert ingredients, making it as powerful as liquid nitroglycerin. The resulting gel could be made cheaper by adding sawdust, black powder, nitrates, or chlorates. Nobel finally decided on a mix of the gel with nitrate and sawdust and called the new product gelignite.

The same method of mixing in cheaper materials such as sawdust and nitrates was applied to dynamite. This is now called *dynamite with an active base*, and has completely superseded diatomaceous dynamite in the United States. To keep the mixture from absorbing moisture, fats, waxes, or paraffin are added to the mix.

The pure gel of guncotton dissolved in nitroglycerin Nobel later patented in 1888 as the propellant *ballistite*, to be used in artillery and small arms. As a smokeless powder, ballistite was an improvement over pure guncotton, and a class of propellants called double-base propellants replaced the (now known as) single-base propellant guncotton. The British invention of the substantially similar *cordite* led to Nobel suing for patent infringement, but he lost the case in the British House of Lords. Cordite was patented by Frederick Augustus Abel (known for making guncotton practical) and Sir James Dewar, after the two had been evaluating Nobel's ballistite and discussing its manufacture with Nobel.

Guncotton is cellulose nitrated in varying amounts. The more the cellulose molecule is nitrated, the less soluble it is in a 50/50 mix of diethyl ether and alcohol. Nobel thought that the less nitrated form would make a better propellant, as it would be less likely to shatter the gun. Abel and Dewar thought the same, and were surprised to find out that the more highly nitrated form (which was known to be more stable against degrading) was not more powerful, and worked as well in guns as the soluble form. The key to the court case ended up resting on Nobel's patent claiming that the soluble form should be used and that the insoluble form (which cordite used) was not suitable.

PICRIC ACID

The Irish chemist Peter Woulfe may have been one of the last practicing alchemists. In his later years he was known as an eccentric, perhaps deranged, person, who attached prayers to his chemical apparatus, and attributed his failures at transmuting base metals to gold and finding the elixir of life to not having performed enough pious and charitable acts. In his more lucid days, however, he was a Fellow of the Royal Society of London and made several important contributions to the study of chemistry. He was a recipient of the prestigious Copley Medal "For His Experiments on the Distillation of Acids, Volatile Alkalies, and Other Substances."

In 1771 he reported in the *Philosophical Transactions* of the Royal Society his experiments with nitric acid on the blue dye indigo (the same dye used for coloring blue jeans):

A Method of dying Wool and Silk of a yellow colour, with Indigo; and also with several other blue and red colouring Substances.

The Saxon blues have been known for some time; and are made by dissolving indigo in oil of vitriol [sulfuric acid], by which means the indigo becomes of a much more lively colour, and is extended to such a degree, that it will go very far in dying.

A receipt for making the best Saxon blue will, I dare say, be agreeable to many; I will, therefore, give the following, which produces a very fine colour, and never fails in success.

Mix 31 of the best powdered indigo, with 34 of oil of vitriol in a glass body or matrass: and digest it for one hour with the heat of boiling water, shaking the mixture at different times; then add 312 of water to it, and stir the whole well, and when grown cold filter it. This produces a very rich deep colour; if a paler blue be required, it may be obtained by the addition of more water. The heat of boiling is sufficient for this operation, and can never spoil the colour; whereas a sand heat, which is commonly used for this purpose, is often found to damage the colour, from its uncertain heat.

Indigo, which has been digested with a large quantity of spirit of wine [ethanol], and then dried, will produce a finer colour than the former, if treated in the same manner, with oil of vitriol.

No one, that I know of, has heretofore made use of the acid of nitre [nitric acid], instead of the acid of vitriol; and it is by means of the former that the yellow colour is obtained: it is nevertheless natural to use it, on account of its known property of making yellow spots, when dropped on any colored cloth.

The acid of salt [hydrochloric acid] does not dissolve indigo, and therefore is of no use in dying.

Receipt for making the yellow dye.

Take $3\frac{1}{2}$ of powdered indigo, and mix it in a high glass vessel, with 32 of strong spirit of nitre, previously diluted with 38 of water; let the mixture stand for a week, then digest it in a sand heat for an hour or more, and add 34 more of water to it; filter the solution, which will be a fine yellow colour.

Strong spirit of nitre is liable to set fire to indigo; and it is on that account that it was diluted with water, as well as to hinder its frothing up. $32\frac{1}{2}$ of strong spirit of nitre will set fire to $3\frac{1}{2}$ of indigo; but, if it be highly concentrated, a less quantity will suffice.

If the indigo be digested for twenty four hours after the spirit of nitre is poured on it, it will froth and boil over; but, after standing a week or less, it has not that property.

One part of solution of indigo in the acid of nitre, mixed with four or five parts of water, will dye silk or cloth of the palest yellow colour, of any shade to the deepest, and that by letting them boil more or less in the colour. The addition of alum is useful, as it makes the colour more lasting; according as the solution boils away, more water must be added.

None of the colour in the operation separates from the water, but what adheres to the silk or cloth; a consequence this colour goes far in dying.

Cochineal, Dutch litmus, orchel, cudbear, and many other colouring substances treated in this manner, will all dye silk and wool a yellow colour.

The indigo which remains undissolved in making Saxon blue, and collected by filtration, if digested with spirit of nitre, dyes silk and wool of all shades of brown inclined to yellow.

Cloth and silk may be dyed green with indigo; but they must first be boiled in the yellow dye, and then in the blue.

The yellow dye Woulfe created is today known as picric acid, or more formally, 2,4,6-trinitrophenol.

Picric acid is a yellow crystalline solid. It dissolves in water, forming a highly acidic solution. It was used as an antiseptic for bandages and as a burn treatment, but its toxicity and tendency to form sensitive explosive salts on contact with metals such as copper, iron, lead, or the calcium in concrete caused problems, and these uses were discontinued.

Picric acid by itself is not a sensitive explosive. However, because it is a strong acid, it easily forms salts with metals, and these salts are often quite sensitive to heat, friction, or percussion. Primary explosives (those that are easily set off) are easily recognized, as during normal handling in the laboratory they make their nature quite clear, often with injuries. Secondary explosives, those that are insensitive to shock and heat, are often not recognized as explosives for some time. Picric acid is one such example.

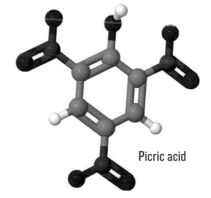

Picric acid

The French chemist Jean-Joseph Welter found the same molecule when he nitrated silk in 1799. He experimented with its explosive properties that year. In 1809 the French chemist Michel Eugène Chevreul demonstrated it was an acid and contained nitrogen. The French chemist

Jean-Baptiste-André Dumas named it picric acid, from the Greek word for bitter. It was not known that Welter's Bitter, as it was called, was the same as Woulfe's yellow dye until the German chemist Justus Freiherr von Liebig experimented with both and pronounced them the same substance. Liebig named it carbazotique acid in 1828. Both names were in use for some time.

In 1842 the French chemist Auguste Laurent made it from phenol (known then as carbolic acid), showed that three of the hydrogen atoms in phenol had been replaced by NO_2 groups, and named it tri-nitrophenic acid. In 1869 he found an easier method of preparing it using phenol sulphonic acid as the base. That compound is made by dissolving phenol in concentrated sulfuric acid.

By 1867 industrial processes were in place for making picric acid, ammonium picrate, and potassium picrate. In that year the Italian chemist Luigi Borlinetto made a mixture of potassium chlorate and picric acid to use as an explosive, but it was too sensitive. It did find use as a flash powder for photography. He went on to try a mix of picric acid with potassium dichromate, which was more powerful than black powder as an explosive but apparently not accepted as a replacement.

In 1873 the German chemist Hermann Sprengel showed (again) that picric acid could be detonated. By 1886 the French had created a bursting charge for artillery shells made of picric acid and nitrocellulose, which they called Melinite. The advantage of picric acid for this use was that it was not so sensitive that the shells would explode inside the cannon from the sudden acceleration of the modern propellants.

The British made an explosive using the same ingredients they called Lyddite, after the town of Lydd, where it was manufactured, and in 1894 the Russian military was also filling shells with picric acid. A related explosive, ammonium picrate, was called Dunnite (or explosive D) by the United States in 1906.

Picric acid was used by British forces retaking the Sudan in 1898, and again in the Second Boer War (1899 to 1902), by both sides in the Russo-Japanese War (1904 to 1905), and by all sides (particularly the French) in World War I (1914 to 1918), although by 1902 Germany had begun to use TNT.

Picric acid caused problems when it reacted with the metal casing of shells, forming metal picrates that are much more sensitive to percussion, friction, and heat. The metal was often protected by tin plating (tin does not react with picric acid) or asphaltum, which helped but could be scratched off in handling. Nothing could be done about the sensitivity of metal picrates to heat, however. On May 1, 1916, a French ammunition factory caught fire, and melted picric acid poured onto the concrete floor, where it reacted with the calcium in the concrete. The heat of the fire detonated the calcium picrate, which then detonated the picric acid. The blast destroyed the factory and killed 170 people.

On December 6, 1917, the French cargo ship SS *Mont-Blanc* left Halifax Harbor in Nova Scotia, fully loaded with explosives for use in the war, mostly picric acid but also TNT, guncotton, and high-octane fuel. In a narrow strait in the harbor, it collided at low speed (about one mile per hour) with an unloaded Norwegian vessel leaving for New York to collect relief materials for Belgium. The collision resulted in a fire aboard the SS *Mont-Blanc*. Twenty minutes later the *Mont-Blanc* exploded in the largest chemical explosion ever seen, releasing the equivalent of 2.9 kilotons of TNT. Structures within a half mile were obliterated, including the entire town of Richmond, killing two thousand people and injuring nine thousand. The explosion caused a tsunami that wiped out a community of indigenous people of the Mi'kmaq tribe.

Halifax explosion

TRINITROTOLUENE

Pierre-Joseph Pelletier was a chemist and director of the School of Pharmacy in Paris in 1837. Famous for earlier work, where he was the first to isolate chlorophyll, quinine, and strychnine, along with several other plant alkaloids, he was a frequent collaborator with the Polish chemist Filip Walter.

In that year (1837) the two were distilling pine oil and isolating its constituents. One of those was a substance new to science, the molecule *toluene*.

Toluene is a good starting point for more complex molecules. It's like a benzene ring with a handle, and the handle helps when adding toluene to other molecules to make whatever the organic chemist is trying to make.

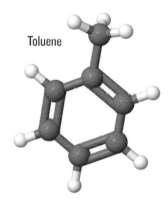

Toluene

In 1863 the German chemist Julius Wilbrand was trying to make dyes. An important class of dyes at the time was the aniline dyes, made from the molecule aniline. The Russian chemist Nikolai Zinin (Alfred Nobel's childhood tutor) had made aniline from nitrobenzene by adding hydrogen in 1842. So, it was natural to suspect that nitrating toluene might make a good start at creating a new type of dye. After all, picric acid was a yellow dye, and using toluene instead of phenol seemed a reasonable substitution.

Adding an NO_2 group to toluene is fairly simple. Adding another NO_2 group is a little bit harder. Adding a third NO_2 group is yet another step. In the second and third steps, the groups can end up on adjacent carbons, or they can be arranged symmetrically to form the molecule 2,4,6-trinitrotoluene, which is the molecule Wilbrand came up with. It makes a nice pale yellow dye. Similar molecules are used today in hair dye products.

It would be natural to investigate the uses of the new molecule as an explosive. A very similar molecule, 2,4,6-trinitrophenol (known as picric acid), was already in use as an explosive, as was trinitroglycerol (another name for nitroglycerin). But trinitrotoluene has a high activation energy, making it difficult to detonate, and it is less powerful than picric acid or nitroglycerin. Moreover, it is toxic.

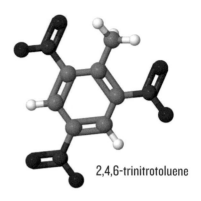

2,4,6-trinitrotoluene

The difficulty in detonating TNT became an advantage when German munitions manufacturers wanted to fill shells with an explosive that would not detonate until after the shell had pierced the armor of enemy defenses. Add to this its insolubility in water, and the ease of melting it (it melts at 80°C without decomposing) to pour into shells and bombs, and TNT's drawbacks start to look like less of a problem.

Still, it was 1902 before German artillery shells were using TNT and 1907 before the British started using them in shells. To make TNT less expensive, and to stretch supplies during wartime, it was mixed with ammonium nitrate to make the explosive amatol.

TNT makes large amounts of black smoke when it detonates because there is not enough oxygen in the molecule to completely oxidize all of the carbon atoms. The result is sooty smoke. Adding ammonium nitrate, an explosive in its own right, corrected the oxygen balance, allowing those carbon atoms to react and add to the explosive force. However, the major benefit of ammonium nitrate was that it was cheap and plentiful. If the amount of TNT in the amatol stayed above 60%, there was little reduction in the power of the explosive, despite the fact that ammonium nitrate is less powerful.

Oxygen balance is the term for indicating how well matched the fuel and oxidizers are in an explosive. If all of the carbon, hydrogen, and metals in a molecule are completely oxidized, then the oxygen balance is zero. TNT has an oxygen balance of −74%, showing that is has insufficient oxygen. Since the remainder of the molecule is mostly carbon, the thick black sooty smoke of a TNT explosion is the result.

When an explosive detonates, one can generalize the reaction products based on what happens to the oxygen. Whatever the actual chain of reactions might be, a rule of thumb is to take the reactants one by one with the following rules:

1. Combine all the nitrogen atoms to form N_2.
2. Then burn all the hydrogen with any oxygen to form H_2O.
3. Next, use whatever oxygen remains to burn carbon to CO (carbon monoxide).
4. If oxygen remains, burn CO to CO_2 (carbon dioxide).
5. Last, any remaining oxygen produces O_2 (molecular oxygen gas).

Ammonium nitrate has an oxygen balance of +20%, so it releases extra oxygen when it explodes. But the velocity of detonation for ammonium nitrate is low, and a mixture of 80% ammonium nitrate with 20% TNT has a detonation velocity of 5,130 meters per second, compared to 6,924 mps for pure TNT. The actual detonation velocity of an explosive varies with the size of the particles, whether any solvents remain in the mix, whether the product is cast or cold pressed, and the size of the charge, among other considerations. The velocities given by one test will differ by as much as 300 mps from another, under different circumstances.

The 80/20 mix of amatol has an oxygen balance of +1. An "ideal" oxygen balance would be zero. When comparing explosive power to pure TNT, however, the 80/20 mix is 30% more powerful. Thus, it is evident that simple oxygen balance is not all there is to know about an explosive.

What balancing the oxygen does for amatol is increase the amount of gas that was produced. So while the brisance (the ability to shatter) is reduced, due to the lower velocity, the ability to move material (the heave) is increased.

In mining, sometimes the rock is hard, requiring a brisant explosive, and sometimes it is softer or porous, in which case an explosive with less shattering ability and more ability to heave the rock out of the way is desired. One can think of brisance and heave in terms of a lever—how hard you have to push can be traded for how far you have to push.

Adding tiny flakes of aluminum to amatol gives the explosive ammonal. The aluminum helps to balance the oxygen, using up the extra oxygen the ammonium nitrate provided to amatol. It also burns exceedingly hot and produces a bright flash at night, helping the artillerymen to adjust their aim. During the day ammonal produces white smoke, which can also be seen from a distance and used for aiming adjustments.

Ammonal was used effectively in World War I. In the Battle of Messines, for example, British engineers tunneled under German front lines and planted over 13 tons of ammonal explosive, which created nineteen large craters and killed an estimated ten thousand German troops.

The relative explosive power is a function of the energy released in the explosion, multiplied by the amount of gas produced, divided by the square of the weight of the explosive. To make comparisons easier, it is now common to compare explosives to TNT. Since the comparison is done by weight, the density of the explosive becomes significant. To equal the force of 1 kilogram of TNT, for example, you would need about 2.4 kilograms of ammonium nitrate.

Some relative effectiveness values for a few explosives:

Ammonium nitrate	0.42
Black powder	0.55
ANFO	0.74
TATP	0.80
50/50 Amatol	0.91
TNT	1.00
80/20 Amatol	1.10
Nitrocellulose	1.10
Picric Acid	1.20
C4	1.34
Nitroglycerin	1.54
RDX	1.60
PETN	1.66
HMX	1.70
Octanitrocubane	2.38

8

TETRYL

In 1877 German chemists Wilhelm Michler and Carl Meyer invented the compound tetryl, or 2,4,6-trinitrophenylmethylnitramine. It took until 1886 before its structure was established by Dutch chemist Karel Hendrik Mertens. Another Dutch chemist, Pieter van Romburgh, later (in 1889) proved the structure by synthesizing it from picryl chloride and potassium methylnitramine.

Tetryl was the first of the nitramine class of high explosives, which includes many of the most explosive compounds currently in use. Others in this class include RDX and HMX, developed much later. Nitramines are characterized by the two nitrogen atoms connected together, with one of them then connected to two oxygen atoms. Tetryl has one nitramine group, shown at the top of the drawing.

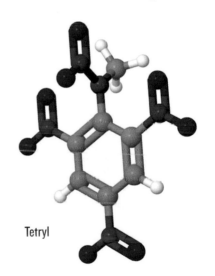

Tetryl

Tetryl is about as sensitive to shock and heat as picric acid, making it a good booster, but generally not as safe as TNT or ammonium picrate. As a booster, it is used in blasting caps, with primary explosives such as mercury fulminate and potassium chlorate to set it off. The tetryl then quite reliably sets off the main charge of TNT or other less sensitive explosive, due to the very high detonation velocity and brisance of tetryl.

Tetryl can be detonated by a spark or by friction but is almost always detonated by a primary explosive in actual use.

Tetryl was used in both world wars as a booster and by itself in some of the smaller caliber shells. It was more expensive than TNT and amatol, but only small quantities were needed since it was not often used as the main charge. Small quantities is a relative term—in World War I the United States alone used 1.5 million pounds up to September 1, 1918, and contracted for 1,432,000 pounds to be delivered by December 31 of that year. Prior to the war the US production of tetryl was negligible.

A mixture of 70% tetryl and 30% TNT is called *tetrytol*. As a booster, tetrytol is safer to use than tetryl, being less sensitive to shock, heat, and friction, and it has the ability to be cast into shaped charges. It was used in burster tubes for nerve gas weapons. It was also less expensive than pure tetryl.

Tetryl was more difficult (and hence more expensive) to produce than other common explosives. Many steps were involved in the production, and many by-products are produced during its manufacture that were difficult to remove. The history of tetryl is thus one of continual search for better and cheaper ways to manufacture it, all the way up to its replacement by RDX and HMX. (Tetryl is no longer manufactured in the United States.)

The long chain of reactions that results in tetryl starts with methyl alcohol. At the time of World War I, methyl alcohol (called wood alcohol) was exclusively made from the distillation of wood. Wartime production of methyl alcohol required more wood than was available, so the supply of methyl alcohol was limited.

The vapor of methyl alcohol was passed over a hot copper mesh, which acted as a catalyst for the production of formaldehyde from the vapor. The reaction is hot enough that the copper mesh stays red hot and only has to be heated at the start.

The formaldehyde is then used to make methylamine (which is also a starting material for RDX and HMX). Methylamine is converted to dimethylaniline by a multistep process and can then be nitrated to form tetryl. The problem with this synthesis (besides the need to start with methyl alcohol) is that many by-products of the reactions accumulate and require difficult and wasteful purification.

A better method, where methyl alcohol is reacted with ammonia by applying heat with a thorium oxide catalyst to form methylamine, which then reacts with dinitrochlorobenzene to form 2,4-dinitromonomethylaniline, was developed. The final product is easily nitrated to tetryl without by-products.

Despite the new method, methyl alcohol was still the starting ingredient and was in short supply during World War I.

PENTAERYTHRITOL TETRANITRATE
(PETN)

Bernhard Christian Gottfried Tollens was a German chemist. In 1891 he and a student, P. Wigand, created the compound pentaerythritol.

Pentaerythritol is similar to glycerin, but instead of having three carbons and three OH groups, it has five carbons arranged symmetrically like a child's toy jack, with four OH groups attached to the four terminal carbons.

Pentaerythritol

In looking for a better explosive than nitroglycerin, it makes sense to start with something similar to glycerin but with more carbon, since nitroglycerin has a positive oxygen balance and the extra fuel would bring the balance closer to zero. Nitroglycerin is what chemists call an ester. An ester is a combination of an acid, such as nitric acid, and an alcohol, which is a carbon chain with an OH group attached. The acid, in this case HNO_3 (nitric acid), loses its hydrogen to the OH group to make HOH (water, or H_2O). The remaining part of the acid attaches to the carbon that used to have the OH group attached.

By slowly pouring pentaerythritol powder into an ice-cold mixture of concentrated sulfuric and nitric acids, four NO_3 groups can be attached to the legs of the molecule. This is exactly what was done at the German munitions factory of Rheinisch-Westfälische Sprengstoff A.G. in Cologne, Germany, in 1894.

PETN is one of the most explosive compounds known. It is 1.66 times more explosive than TNT.

In 1912 the German government patented a better way to produce it, just in time for use in World War I.

PETN degrades more slowly than nitroglycerin or nitrocellulose but it is still less stable than TNT. It is less shock sensitive than nitroglycerin but more sensitive than TNT. For these reasons it is often mixed with TNT or other explosives, resulting in a mix that has the high brisance of PETN but is less sensitive and stores better.

PETN is the ingredient used in primer cord, an explosive string that is used to set off other explosive charges. Going by the name of *primacord*, or *detcord*, it is a plastic tube full of PETN. By connecting separately spaced charges with detcord, the extremely high detonation velocity of the PETN leads to near-simultaneous detonation of all the charges.

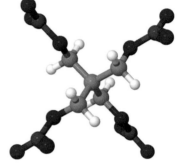

Pentaerythritol tetranitrate

A few wraps of detcord around a tree can cut it down much faster than a chainsaw, and it is used this way militarily by engineering teams to clear a path for mechanized equipment. In rubberized sheet form with nitrocellulose, PETN makes up Primasheet and Detasheet. These are used for explosive welding applications, since they can be wrapped around the objects to be welded together.

A mixture of 50% PETN and 50% TNT is called *Pentolite*. Pentolite can be cast like TNT, and is more stable than pure PETN, while still having a detonation velocity between 7,400 and 7,800 meters per second. It is used as a booster explosive for harder to detonate explosives like ammonium nitrate–based mixtures.

PETN is often *phlegmatized* (made less sensitive by adding less explosive or nonexplosive materials) by paraffin or wax, or by polymers such as rubber or silicone to make plastic explosives. In this form it can be used in some of the smaller caliber artillery shells.

Because PETN has a low vapor pressure (it does not vaporize easily), it is more difficult to detect by bomb detector equipment or by bomb sniffing dogs. Combined with its high power and ease of manufacture, this has made it a favorite of several infamous terror plots.

Richard Reid, the so-called "shoe bomber," tried to use PETN detonated by TATP to bring down American Airlines Flight 63. An al-Qaeda suicide plot to assassinate Saudi Prince Muhammad bin Nayef almost succeeded with a bomb inserted in the rectum of the suicide bomber. He successfully detonated the bomb, but the target survived. The Nigerian "underwear bomber" tried unsuccessfully to detonate a PETN bomb sewn into his underwear by injecting it with a syringe of liquid believed to be nitroglycerin. A small fire resulted instead of the planned detonation. PETN printer cartridges were found on a Dubai flight, thanks to intelligence work (not by bomb detection). Had they detonated, they would have been more than powerful enough to down the plane.

Like nitroglycerin, PETN is a vasodilator, and can be used to treat angina. In nearly pure form in the drug Lentonitrat, it can be sprayed under the tongue to relieve symptoms and quickly lower blood pressure.

CYCLOTRIMETHYLENETRINITRAMINE (RDX)

The name RDX stands for Research Department eXplosive. This was a code name for the explosive known in the UK as cyclonite, in Germany and Russia as hexogen, and in Italy as T4.

RDX was patented in 1898 by the German chemist Georg Friedrich Henning. He prepared it by nitrating hexamine nitrate. His original patent (German patent 104280) mentioned medical uses, but subsequent patents suggested its use in smokeless propellants.

The original patent is short enough to show here in its entirety:

Cyclotrimethylenetrinitramine (RDX)

Patent issued in the German Republic on July 15, 1898.
Dr. G. F. HENNING, Berlin, Germany
Process for Making a Nitro compound from Hexamethylenetetramine

Hexamethylenetetramine cannot be nitrated using the commonly known nitration methods; the tetramine compound decomposes into ammonia and formaldehyde. It is, however, possible, using a bypass to obtain a nitro compound, which has remarkable properties. Dissolving hexamethylenetetramine in 3 parts of water and cooled to 0° C, adding this solution to cooled concentrated nitric acid, very soon the crystallization of nitric acidic hexamethylenete-tramine begins. This salt must be washed with ice water and cold ethanol and dried immediately. Otherwise, it decomposes forming formaldehyde.

For making the nitro compound 10 parts of the thoroughly dried salt is added in small portions to 50 parts concentrated nitric acid which is cooled to −5°C and has a specific weight of 1.52, after each addition waiting for ceasing the mild foaming caused by the reaction, before added a new portion. After adding the entire amount of the nitrate, the beaker containing the nitrate is allowed to stand in the cooling mixture 30 minutes and pouring the acid in a small stream in ice water. Immediately the precipitation of the colorless crystals begins which are dried in air after it was thoroughly washed with water.

The yield is such that the amount of nitro compound made as the amount of hexamethylenetetramine was used. The previously unknown compound melts at 200° C under decomposition; its composition is as follows:

38.34% nitrogen
16.45% carbon
2.81% hydrogen
42.40% oxygen

The simplest net formula is $C_3N_6H_6O_6$.

The compound does not dissolve in water; it dissolves slowly in boiling alcohol, however, it dissolves easily in acetone in which it can be made in any crystal sizes. The solution in acetone reacts acidic. The compound is also easily dissoluble in concentrated acetic acid and precipitates forming crystals when diluted with water. The nitro compound does not dissolve in diluted acids, as well as in diluted alkali. If heated in potassium or sodium hydroxide it decomposes, and the decomposition continues without any further heating once it is initiated. Among the decomposition products was nitrogen and oxygen. When rapidly heated the compound explodes with a bang without any residue; it is not sensitive against blow and stroke, its storage stability is unlimited. The novel nitro compound does not have an odor and neither taste, however, it forms easily formaldehyde in the presence of reduction and oxidation agents; the nitro compound has technical applications and serves as starting material for medicinal products.

Claim:
A process for making a nitro compound from hexamethylenetetramine by adding nitric acidic hexamethylenetetramine in cold concentrated nitric acid.

In 1920 Austrian (later a citizen of Germany) Edmund von Herz filed a patent (US1402693) on an improved method for making it.

During World War II several new methods for making RDX were developed, several in Germany, and more each in the United States, Canada, and the UK.

Specifically, the American chemist Werner E. Bachmann and his newly minted PhD student John Clark Sheehan developed a method involving the nitration of hexamethylenetetramine.

Sheehan recalls wearing the standard lab coat and safety glasses, but also a heavy towel wrapped around his neck to protect it from flying glass. They made the explosive at a kilogram scale, and the process was later scaled up further by the company Tennessee Eastman (part of Eastman Kodak) for the war effort. Because the Bachmann process was a continuous flow method, not the batch methods in use elsewhere, it could keep up with the war demands more easily.

During the war Britain was finding it difficult to destroy German U-boats by air using only TNT-based explosives. RDX, being much more powerful, was used in a mixture containing TNT and aluminum powder, resulting in half-again more powerful depth charges that could be dropped by aircraft after sighting an enemy submarine. The mixture, called Torpex, was also used to burst dams, in a type of bomb that would skip over the water to avoid torpedo nets before hitting the dam and submerging to detonate far underwater and produce the highest shockwave.

In the United States wartime production of RDX amounted to 15,000 tons a month. In Germany another 7,000 tons a month were produced. While more expensive to manufacture than TNT, RDX had twice the explosive power as the same volume of TNT.

RDX has a very high detonation velocity—the densest form detonates at 8,750 meters per second.

RDX became a mainstay in military explosives, almost as much as TNT had been before it. It was mixed in a variety of forms under a number of names:

- Torpex—42% RDX, 40% TNT, 18% powdered aluminum
- Semtex—50% RDX, 50% PETN
- Cyclotol—A castable explosive mixture of TNT and RDX in various proportions
- Composition A—RDX and wax, used to desensitize it and make a moldable explosive
- Composition B—60% RDX and 40% TNT, plus some desensitizing wax
- Composition C—91% RDX mixed with a plasticizer, a binder, and motor oil
- PBX—Polymer-bonded explosive, a huge number of plastic explosive formulas

Composition C was later subdivided into several types, C-2, C-3, and C-4.

The starting point in the synthesis of RDX is methyl alcohol. As with tetryl, methyl alcohol (produced by the distillation of wood) was a scarce commodity during World War I, limiting the availability of both tetryl and RDX.

In 1921 the French chemist Georges Patart patented a method of producing methyl alcohol from carbon monoxide and hydrogen using catalysts and high pressure. Improved catalysts developed in Germany allowed the industrial production of large quantities of methyl alcohol, thereby destroying the wood distillation industry.

By the late 1920s methyl alcohol was being produced in the United States by yet newer methods, starting with coal as the base for both the carbon monoxide and hydrogen, made by passing steam over hot coke (a by-product of the distillation of coal).

By adding extra hydrogen (from reacting steam with hot iron), methyl alcohol could be made from carbon dioxide instead of carbon monoxide, and since large amounts of carbon dioxide were available from industrial fermentation plants, the price of the alcohol went down, and the availability was limitless.

Reacting methyl alcohol with a hot copper screen as a catalyst produces formaldehyde, a starting point for the production of RDX.

After 1926 RDX could be made from nothing other than coal, water, and air. With virtually unlimited access to the starting materials, only the lengthy synthesis process made tetryl and RDX more expensive than the less powerful TNT.

CYCLOTETRAMETHYLENE TETRANITRAMINE (HMX)

When Bachmann and Sheehan were developing their method of manufacturing RDX, Sheehan isolated another explosive compound during the purification process. This turned out to be an even more explosive, denser molecule, cyclotetramethylene tetranitramine, now known as HMX. About 8% to 10% of the yield from the Bachmann RDX production method is HMX.

In the place of RDX's six-sided molecule is an eight-sided molecule with the same decorating nitro groups attached. Instead of RDX's three NO_3 groups, it has four, as tetryl does.

The extra nitrogen atoms and the extra nitro group give HMX a detonation velocity of 9,100 meters per second (compare to 8,750 for RDX or 7,570 for tetryl).

In Germany and Russia HMX is called octogen (recall that RDX is called hexogen there). The name HMX is an obvious derivative of its close relation RDX, and there are many guesses as to what the acronym means (High Melting eXplosive, High-velocity Military eXplosive, etc.), but no actual records are available. It may have simply been a code word.

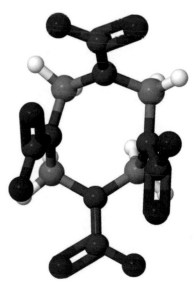

Cyclotetramethylene tetranitramine (HMX)

Initially HMX was simply extracted as a by-product during the manufacture of RDX by the Bachmann method.

In 1961 Canadian chemist Jean-Paul Picard patented a method of making HMX directly from hexamethylenetetramine (US2983725). Previous methods had yields of only 50%. The new method had a yield of 85%, with purity over 90%. A drawback of the Picard method is that it is a multistage process with aging intervals, which makes it slow.

In 1964 the Indian chemists H. K. Acharya and R. T. Limaye developed a single-stage process, bringing the cost down considerably.

HMX is more stable than RDX. It ignites at a higher temperature (335°C instead of 260°C), and it has the chemical stability of TNT or picric acid. It has a higher detonation velocity, and in Trauzl lead-block expansion tests, it expands a cavity in a lead block by 48 cubic centimeters per gram to RDX's 45.

HMX is used where its high power outweighs its expense (about $100 per kilogram). In missile warheads, for example, a smaller charge of a more powerful explosive allows the missile to travel faster or have a longer range. It is also used in shaped charges to penetrate armor and defenses, where a less brisant explosive would fail.

A mixture of 70% to 75% HMX with TNT goes by the name Octol. Its advantage is that it is castable and yet retains the high brisance of HMX. Another mix, called OKFOL, is 95% HMX desensitized with 5% wax, which also makes it formable, although the detonation velocity falls to 8,670 meters per second.

HEXANITROHEXAAZAISOWURZITANE
(HNIW OR CL-20)

In a corner of California's Mojave Desert sits a naval air base called China Lake. It is the size of that key metaphor for the sizes of large areas—the state of Rhode Island. Located in a desert many miles from the dense population areas, it is the perfect place to develop and test the latest high explosive compounds.

More commonly called CL-20 for some reason (the CL is for China Lake, of course), the cage-like molecule hexanitrohexaazaisowurzitane is more powerful than HMX (9,660 meters per second detonation velocity to HMX's 9,100 mps) and denser, fully twice the density of water. Developed in 1986 by Arnold Neilsen, it is one of the densest organic molecules known and has the highest energy content.

Two things give CL-20 its advantage over HMX, the previous record holder for most powerful explosive. First, it has a higher oxygen balance, so more of the carbon atoms are oxidized. In addition, the cage structure has several strained bonds. Bonds that normally would end up at 120 degrees are pushed and pulled into different angles. Like bending a bow to shoot an arrow, these strained bonds store extra energy in the molecule, above that delivered when the molecule rearranges into the reaction prod-

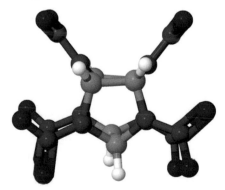

Hexanitrohexaazaisowurtzitane (CL-20)

ucts. These strained bonds also make the compound denser, increasing its performance.

CL-20 is a little more sensitive than HMX (closer to PETN, one of the least stable of the modern military explosives), but it is most commonly bonded to rubbery polymers in a polymer-bonded explosive, which helps to manage the sensitivity. In mixtures with HMX and TNT, it improves the performance of the other explosives and gains some of their insensitivity.

Because of the complexity of the molecule, a view from another angle is interesting.

One of the ways to reduce sensitivity to friction and impact or shock is to cocrystallize the molecule with another, less sensitive, high explosive. In a cocrystal, molecules of one substance combine with those of another in integer combinations

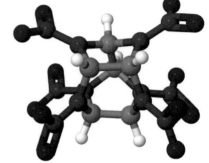

Hexanitrohexaazaisowurtzitane (CL-20)

such as 1:1 or 2:1. In a cocrystal of CL-20 and HMX, there are two CL-20 molecules to each HMX molecule in a nice regular crystal structure. This reduces the detonation velocity to 9,470 meters per second, what would be expected as an average of the two explosives. However, the surprise was that the sensitivity was the same as plain HMX. The cocrystallization seems to have made the CL-20 more stable, and the HMX becomes the limiting factor. The increased density of the result makes the mixture 20% more powerful than HMX.

As a high-energy molecule, it finds uses besides as an explosive. It has been developed as an efficient solid rocket fuel with the advantage that (because of its oxygen balance) it produces less smoke than the mixtures it replaces. Its high density also means more propellant can be packed into the missile, giving it more range and speed. As unmanned aerial vehicles get smaller, there is a premium put on the size of the arms they carry.

The current cost (in 2015) is over a thousand dollars per kilogram. This is expected to drop tenfold, into the HMX price range, when production scales up.

TRIAMINOTRINITROBENZENE (TATB)

TATB is a benzene ring with alternating nitro and amino groups attached around it. As an explosive, it has a detonation velocity of 7,350 meters per second, less than RDX (8,750 mps) but more than TNT (6,900 mps).

The main distinguishing feature of TATB is its insensitivity to shock, impact, vibration, or flame. It is extremely difficult to accidentally detonate. This is a big concern with nuclear weapons, which are activated by high explosives—an accidental nuclear explosion could be somewhat embarrassing.

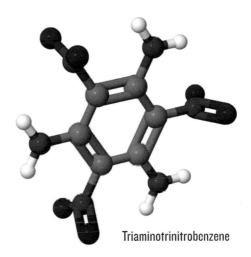

Triaminotrinitrobenzene

Explosives that might be carried by aircraft need to be able to hit the ground without going off in the event of a crash.

TATB is sometimes mixed with other explosives (such as 15% HMX in the plastic-bonded explosive PBX-9503), but usually it is used alone, to make the most out of its insensitivity to accidental detonation.

TATB, first tested in 1955 and 1956, is the first in a class of explosives selected for their insensitivity. Others in this class include FOX-7.

When using a more sensitive explosive, such as TNT, RDX, or HMX, a detonating cord can be sufficient to set off the device. TATB will not detonate reliably from detonating cord, so a more sensitive secondary explosive (such as HMX) is used, and the TATB becomes a tertiary explosive.

TATB is a very flat molecule, similar to graphite. This gives it a lubricating effect (lubricity) that makes it easy to press into molds while retaining the high density needed for a high explosive. When coupled with a fluoropolymer (such as Teflon) in a polymer-bonded explosive, this effect is increased.

POLYMER-BONDED EXPLOSIVES (PBX)

Some explosives, such as TNT, can be safely melted and cast into artillery shells and missile warheads. Other explosives need a little help, in the form of a small amount of an additive (generally less than 10%) that forms either a rubbery elastic material or a hard machinable material when mixed with the explosive.

With an explosive such as HMX, an elastomer can make the final product rubbery and reduce the likelihood of an unplanned explosion due to shock. Fluorinated polymers (such as Teflon) are inert and thus preferred for explosives that are chemically reactive. They can be solid in form or they can be elastomeric. Solid forms require a less sensitive explosive since the rigidity of the plastic may enhance the sensitivity to shock.

Some binders are explosive themselves, such as those made by nitrating a polymer. They thus serve two purposes: they bind the ingredients into shape and contribute to the force of the explosion. Inert binders add mass that does not contribute to the explosion.

Plastic-bonded explosives find use in many nuclear weapon detonation systems. An example is PBX 9407, a mix of 95% HMX, 2.5% Estane rubber, and 2.5% BDNPA-F (a nitrated plasticizer made from BDNPA and BDNPF).

bis(2,2-dinitropropyl)acetal (BDNPA)

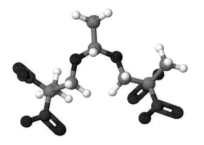

bis(2,2-dinitropropyl)formal (BDNPAF)

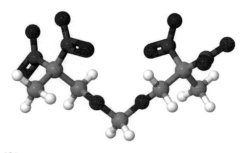

Other PBX formulas use RDX, PETN, or TATB as the explosive and binders like polystyrene plastic, nitrocellulose, polyurethane rubber, hydroxyl-terminated polybutadiene rubber, or Viton fluoropolymer elastomer.

Detonating cord made from PBX is less sensitive and more waterproof than those filled with pure PETN. Torpedo warheads and other underwater weapons also benefit from the water resistance of PBX explosives, which are also much less sensitive to shock and impact than previous charges. They are also better explosives than the TNT and RDX they replaced.

One criterion for selecting an elastomeric binder is its glass transition temperature. It is no longer elastic and becomes brittle below this temperature. This greatly reduces its ability to protect the explosive from shock. If the weapons are to be used or stored in cold weather, this can be a serious issue.

In shaped charges the binder is responsible for keeping the carefully engineered form of the explosive in the right shape. If a binder is not resistant to fatigue, the shaped charge will not perform properly. Flaws and cracks also leave the explosive vulnerable to shock detonation or fragmentation.

In factories with equipment designed for melt-casting TNT, a binder that has the same properties can save money by saving the need for retooling. Likewise, equipment designed for extruding or cold-pressing TNT can still be used if the right binder is chosen in a PBX mix.

As late as 1960 the search for a castable explosive that could withstand 500°C (e.g., in a plane crash) had led to *compressible* compositions based on diaminotrinitrobenzene (DATB) bonded with fluorocarbon binders and pressed into shape. This mixture could survive as high as 400°C. But *castable* explosives close to this temperature

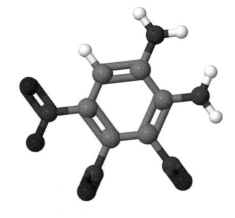

Diaminotrinitrobenzene (DATB)

range were generally using HMX (melting point 286°C), and TNT melts well below these temperatures (80°C).

Casting an explosive like TATB in a thermosetting binder such as epoxy or polyester resin (or a more heat-tolerant binder, such as silicone or fluoropolymers) was considered in 1967, but the characteristics that make TATB insensitive to detonation also make it insoluble in almost all solvents. As it is manufactured in the form of a fine powder and cannot be dissolved and recrystallized in larger form, the loading density (the ratio of explosive to binder) would be too low, lowering the detonation velocity below that of TNT.

More recent research has centered on energetic polymers (which is to say, explosives that can polymerize into long rubbery or plastic-like chains). BDNPA and BDNPF are two examples already mentioned, but more recently molecules like poly(glycidyl nitrate), called polyGLYN, glycidyl azide polymer, called GAP, and poly(3-nitratomethyl-3-methyloxetane), called polyNIMMO are being used.

Glycidyl nitrate (polyGLYN)

Glycidyl azide (GAP)

A mixed polymer of 3-azidomethyl-3-methyloxethane (called AMMO) and 3,3-bis(azidomethyl)oxetane (called BAMO) is amusingly known as AMMO/BAMO, and is used in rocket propellants and as an energetic binder for FOX-7 explosives.

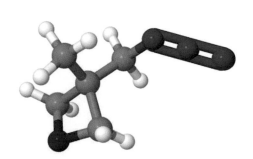

3-azidomethyl-3-methyloxethane
(AMMO)

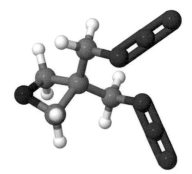

3,3-bis(azidomethyl)oxetane (BAMO)

Polymer-bonded explosives are frequently given names that reflect where they were developed. Those developed at the Lawrence Livermore National Laboratory have designations that start with LX- followed by a number, such as LX-09, LX-14, or LX-14-0. Those developed at the Los Alamos National Laboratory start with PBX and have four-digit numbers starting with 9, such as PBX-9502. Sometimes in the literature, the hyphen is omitted or replaced by a space. Polymer-bonded explosives developed by the US Navy have names that start with PBXN- and a number, such as PBXN-7. Other designations may have the initials of the laboratory in the name, such as PBXIH-18 (developed at the US Naval Surface Warfare Center in Indian Head, Maryland), or PBXW-115 (developed at the US Naval Surface Warfare Center, White Oak, Maryland). Some extrudable explosives have an XTX- designation, such as XTX-8003.

TESTING EXPLOSIVES

As new explosive compounds were created, it became evident that there was a need for a way to compare them. Explosives differ in brisance (the ability to shatter something); heave (the amount of material an explosive can move); sensitivity to heat, shock, or friction; and things like melting point, detonation velocity, etc.

Developing standards for these aspects of explosives allowed customers to specify performance benchmarks and vendors to advertise the advantages of their wares.

Isidor Trauzl, born in Hungary, was an Austrian army captain, chemist, and explosives expert. In 1870 he established the first Austrian dynamite factory with Alfred Nobel and led the company in the years 1882 to 1892.

Trauzl is remembered today for the eponymous Trauzl lead-block test for explosives. It is a way of comparing two explosive mixtures or compounds by measuring how much they can expand a cavity in a lead block.

In the test a cylindrical block of lead, 20 centimeters high and 20 centimeters in diameter, has a hole drilled in the center of the flat face that is 2.5 centimeters in diameter and 12.5 centimeters deep. Into the hole is placed 10 grams of the test explosive and a standard number 8 blasting cap. The hole is then filled with fine sand (tamped) and the explosive is set off.

The hole in the lead block expands due to the explosion. Water is poured into the cavity, and then the water is poured into a graduated cylinder to measure the volume. The original volume of the hole is subtracted to get the volume of the expansion. Some part of the expansion is due to the blasting cap, and this is subtracted from the total as well, as determined by setting off the blasting cap with a nonexplosive 10 grams of dummy material in a similar block.

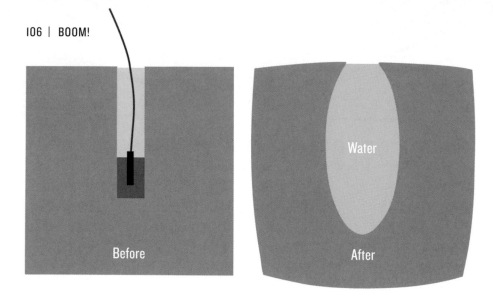

Some part of the explosive energy goes into heating the lead block and the sand. Some part goes into blowing away the sand. Some part escapes as the heat of the gases after they leave the block. So what the test measures is not the total energy of the explosive but the work it does on the lead block.

Frequently the volumes will be standardized to a particular explosive, such as TNT or amatol, and the results given in percent of TNT. Both are shown below.

Nitroglycol	610	203.33%
Methyl nitrate	600	200.00%
Blasting gelatin	600	200.00%
Nitroglycerin	530	176.67%
PETN	520	173.33%
RDX	483	161.00%
Nitromethane	458	152.67%
Gelignite	430	143.33%
Ethyl nitrate	422	140.67%
Tetryl	410	136.67%
Nitrocellulose	373	124.33%
ANFO	316	105.33%
Picric acid	315	105.00%

Trinitroaniline	311	103.67%
Trinitrotoluene	300	100.00%
Urea nitrate	272	90.67%
Guanidinium nitrate	240	80.00%
Ammonium nitrate	178	59.33%

Other tests followed. The steel dent test measures the depth of a dent made in a steel plate by 10 grams of explosive placed in a strong steel cylinder sitting on top of the steel plate.

The sand crush test is done by sifting sand through a sieve, and retaining the sand that did not pass through. A small charge of the explosive is then detonated in the sand. The sand is screened once more, this time measuring the weight of the sand that has been crushed enough to fit through the sieve. This is designed to be a measure of brisance. Results are often normalized to a standard explosive such as TNT.

The ballistic mortar test measures the swing in a pendulum holding a heavy short-nosed mortar. The similar ballistic pendulum test measures the swing of a pendulum when the explosive is detonated next to it. Both tests are prone to experimental variables that are difficult to remove or account for, making comparisons from different laboratories problematic.

In the cylinder test, high-speed photography is used to measure the radial velocity of a metal cylinder filled with the explosive. This measures to some degree the detonation velocity and the effects of the explosive on the metal case.

The airblast test detonates the explosive in the open, and sensors record the air pressure at different distances from the explosion.

In the drophammer test, 35 milligrams of the explosive is placed on the anvil of the drophammer apparatus. A striker is then placed on the explosive. A 2.5-kilogram weight is then dropped from different heights onto the striker. A microphone records the sound and detects an explosion by comparing the sound to the sound of the machine when no explosive is in it. The height that will detonate 50% of the samples is designated the DH_{50} value for that explosive. Often the peak height at which no explosions occur is also recorded.

To prevent rebound the weight is sometimes made as a hollow steel container filled with lead shot. The device is recalibrated every twenty tests by testing with PETN and Composition B.

The drophammer test is a test of the sensitivity of the explosive to impact. For primary explosives, too low or too high a value will be reason to disqualify the compound. For secondary explosives, high values mean the compound is safer to use but may require booster explosives (such as PETN) in the primer cap.

Some typical results are shown below.

PETN	15±3
HMX	31±4
RDX	37±6
Composition B-3	50±4
TNT	61±9

The BAM Friction Sensitivity Test Machine was originally built by the German Bundesanstalt für Materialprüfung, which is how it got the name that sounds so appropriate in English—the BAM test.

The device holds a movable porcelain plate against a fixed porcelain pin. A sample of explosive is placed on the porcelain plate next to the pin. A lever with weights presses down on the pin, allowing masses from 0.5 kilograms to 36 kilograms to press the pin against the plate. The machine slides the plate and sample under the pin a distance of 1 centimeter.

Successive weights are added until the explosive detonates from the friction. The lightest weight that will cause a detonation is the value used. Thus, lower values mean the explosive is more sensitive to friction. A "Go" reaction is anything that can be heard, seen, or smelled. Ten tests are done, and when one in ten tests results in a reaction, that weight is used as the value for that explosive.

Tests are reported either as 1/10 (meaning one explosion out of ten tests) and a weight, or the value 0/10 and a weight of 36 kilograms, indicating the explosive never exploded at the maximum weight.

Some typical results:

ANFO	0/10 36 kg
Black powder	0/10 36 kg
CL-20	1/10 6.4 kg
Composition B	1/10 4.8 kg
HMX	1/10 11.6 kg
Nitrocellulose	1/10 12.0 kg
RDX	1/10 12.4 kg
PETN	1/10 6.4 kg
TNAZ	1/10 11.6 kg

The Gap Test (or Zero Gap) places the sample in a steel tube in front of a steel "witness plate" and sets it off with a Pentolite booster charge of known strength. Changes (damage) to the tube and the witness plate are noted.

There are a number of tests that measure thermal stability. One such test is the Simulated Bulk Auto-ignition Temperature or SBAT test. Samples are put into test tubes with thermocouples to monitor their temperatures. All of the samples are then put into a heated metal bath kept at a constant temperature. The temperatures are recorded as the samples heat up. Active samples heat up more than inactive reference samples.

This tests more than just the autoignition time (how long it takes the compound to explode or burn at a particular temperature). By monitoring the temperature over time, changes in phase (solid to liquid, liquid to gas) can be tracked. Endothermic mixtures (which heat more slowly than inactive reference materials) and exothermic mixtures (those that produce heat) can be distinguished.

Similar tests heat the sample in a sealed steel tube at a constant rate (e.g., 15°C per minute) until it ignites. Fragmentation of the steel tube is noted. Tests such as these are called "cook-off" tests. Some example data from such a test:

PBXN-109	5.12 hours	169.8°C	1 fragment
LX-10 sample 1	15.82 hours	205.3°C	9 fragments
LX-10 sample 2	13.23 hours	197.1°C	7 fragments

The first LX-10 sample in that test was given 10% air headroom (called ullage). The second sample was the same material but without the extra space.

During these tests observers note any discoloration, melting, outgassing, or smoking.

Still other tests are conducted with similar apparatus to monitor how a material behaves when in contact with a test material, such as a prospective packaging or weapon material or transport vehicle material.

Many of these tests are calibrated using some standard explosive and then reported as a percent of the standard. For many years TNT was the standard but more recently RDX has been used. This metric is reported as the Figure of Insensitivity. RDX itself has a Figure of Insensitivity of 80 on the TNT scale (where TNT is 100).

To measure detonation velocity a number of approaches have been used. In one method a piece of resistance wire is placed inside an aluminum tube set next to the explosive. When a length of explosive placed alongside the tube is detonated from one end, the resistance decreases as the detonation front crushes the aluminum tube against the resistance wire from one end of the tube to another. An oscilloscope is used to record the resistance change over time.

A more modern method uses optical fibers placed a known distance apart on a length of explosive. The charge is detonated from one end, and the time between flashes seen by the optical fibers is noted. Older forms used electrical sensors instead of fiber optic cable.

Detonation velocities are often given in millimeters per microsecond, since the equipment measures millimeter lengths and microsecond or nanosecond times, but this gives the same number as kilometers per second (just multiply each by one million).

An old method, the Dautriche method, uses detonation cord placed from one end to the other of a length of the explosive to be measured. The middle of the cord is placed on an aluminum or lead plate. The explosive is detonated from one end, and the detonation cord starts detonating at that end. The detonation wave in the sample under test then reaches the other end and starts that end of the detonation cord exploding. The place where the two racing detonation waves in the detonation cord meet is recorded by the metal plate. Knowing the speed of detonation of the detonation cord allows the tester to calculate the speed of the wave in the sample under test.

High-speed photography of the explosion can give a direct measurement of the detonation velocity.

Sandia National Laboratories developed a system where a coaxial cable (a wire inside a tube) is connected as a delay element in an electronic oscillator. As the cable was crushed by the detonation shock wave, the frequency of the oscillator increased (because the delay is shorter as the tube is collapsed). This was used to measure the shock wave velocity in nuclear weapons tests.

A similar idea is behind another type of sensor. In a coaxial cable any pulse reaching the open end of the cable is reflected back. An instrument sends a pulse down the cable and times how long it takes to see the echo. As the cable gets shorter, less time is needed for the echo to return. This does not depend on the cable collapsing and shorting out. Instead, the cable is progressively disintegrated. At the detonation front, the hot plasma conducts electricity, so it looks like a short circuit. But since a short circuit or an open end both reflect the pulse, this method is very reliable.

Both of these methods, like the resistance wire and photographic methods, give a graph of the detonation velocity over time, instead of a single measurement.

Detonation velocity is affected by the confinement (and thus pressure) of the explosive. Some of the above methods can be used to measure detonation velocities inside boreholes in rock, to get a more accurate measurement of real-world conditions than laboratory testing can give. The detonator is placed at the bottom of the borehole, connected to wires leading up to the surface. Probes—optical fibers, coaxial cables, or electrical sensors—are placed at locations along the borehole, with cables leading to measuring equipment at the surface.

Resistance wire sensors can be as simple as insulated resistance wire twisted together. The detonation wave plasma shorts them out as it travels up the borehole. Equipment at the surface records resistance changes over time. The slope of the recorded line is the detonation velocity.

The detonation velocity is one of the tests that provides a consistent dataset to high resolution. Combined with the density of the explosive (a measurement very easy to make), the so-called CJ pressure (Chapman-Jouguet pressure) of the detonation can be calculated. This value (the CJ pressure) is one of the better metrics to use when comparing explosives, so many lists of explosives give both the density and the velocity of detonation (VOD).

Gap tests are used to tell how easily an explosion in one munition will cause an explosion in a nearby munition. Knowing this information allows supplies to be stacked at a sufficient distance that an accident in one will not cause the rest to be lost. Since space is often at a premium (especially at sea or in the air), knowing how closely munitions can be stored is important. In laying a minefield, if the explosives are too close to one another, all of them might go off when only one is triggered. This is not considered optimal unless you are on the other side engaging in mine-clearing operations.

In civilian mining operations the separation of blast holes from one another must be carefully considered so as to reduce the likelihood of sympathetic detonations. The bores need to be close to one another to properly fracture the rock but not so close that they set one another off. Each blast should be initiated by a blasting cap and nothing else, or excessive ground shaking and ejected material can result. Relief holes can be bored between explosive-filled holes to hinder the detonation wave propagation through the rock.

In the gap test an explosive, called the donor, is detonated next to the explosive under test, called the receptor. A gap between the explosives can be air or some material such as metal or concrete that is being tested for resistance to sympathetic detonations. The distance between the explosives when 50% of the receptor charges detonate is the figure of merit.

INSENSITIVE EXPLOSIVES

In the explosion at Halifax, Nova Scotia, in 1917, the SS *Mont-Blanc* was blown apart after a fire caused her picric acid cargo to melt and then detonate. In numerous naval engagements it has not been a direct hit below the waterline that sinks the ship, but a fire that detonates explosives in the magazine that blows the ship to pieces.

On October 4, 1918, the T. A. Gillespie plant in Morgan, New Jersey, had an accidental explosion. (It was a plant for loading explosive shells.) That explosion caused a fire that triggered numerous explosions over three days. Sixty-two thousand people were evacuated from nearby towns. The damage was estimated at 6 kilotons of TNT explosive, enough to supply the Western Front for six months. Windows were broken as far away as

25 miles. Explosive ordnance from the explosion was scattered for over a mile and was still being found as late as 2007.

On July 17, 1944, munitions being loaded at Port Chicago, California, exploded, destroying the ship, the docks, and killing 320 people.

The list of unintended munitions explosions goes on for pages. Smaller explosions have been innumerable—shells exploding inside large caliber navy guns, explosions in gun turrets, fires on flight decks of aircraft carriers setting off munitions in planes or belowdecks.

In civilian applications, such as mining, well drilling, and oil exploration, explosives are handled and transported daily in close proximity to populations.

The need for safer explosives has led to a class of compounds designed to be difficult to detonate. This effort has led to new explosive molecules like TATB and to new ways to make older explosive molecules less sensitive, like bonding them with rubber or plastic or adding compounds to the mix that desensitize them.

Not all polymer-bonded explosives are less sensitive. PBXN-101, for example, is HMX in a binder of polyester resin. The thermosetting plastic is brittle and very sensitive to shock, bullets, and heat. A later PBX, called PBXN-106, made from RDX and polyurethane rubber, was much less sensitive to shock and fire and was selected for use in naval shells after some disastrous in-bore explosions of older explosives.

After 1970, when accidental and fire-caused explosions during the Vietnam War brought the problems to a new focus, work began in earnest on developing less sensitive explosives.

In some of the first tests of rubber-based PBXs, warheads made with the standard Composition B explosive (60/40 RDX and TNT) and warheads loaded with PBXN-107 (86% RDX in polyacrylate binder) were both exposed to fire from aviation fuel. The Composition B warheads exploded. The PBXN-107 warheads merely split open and burned.

Besides binding with polymers, another way to make explosives like RDX and HMX less sensitive is to reduce the particle size. In tests with mixtures containing micron-sized particles compared to nanometer-sized particles, the nanometer sizes were roughly half as sensitive to friction and impact as the micron sized.

One of the earliest insensitive explosives developed was ammonium picrate, known as Dunnite, or Explosive D. Developed in 1906 by US Army Major Beverly W. Dunn, it is a salt of picric acid. Unlike the metal salts of picric acid, which are generally much more sensitive to heat and shock than the acid itself, the ammonium salt is much less sensitive.

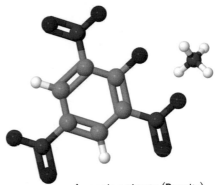

Ammonium picrate (Dunnite)

Dunnite's insensitivity led to its use in armor-piercing shells, as it could withstand the shock of penetrating steel ship hulls, exploding inside the ship on a fuse. In this way it echoed TNT's replacement of picric acid for that same purpose earlier, when armor was lighter.

Ammonium picrate is considerably less sensitive to impact than TNT, although it explodes at 318°C, whereas TNT detonates at 457°C. Like picric acid, ammonium picrate can form metal picrate salts in contact with iron, copper, lead, and other metals, which are extremely sensitive to shock and heat.

Insensitive explosives require powerful explosions to set off all of the charge. When mercury fulminate or lead azide is used to detonate ammonium picrate, not all of the explosive actually detonates. A more powerful initiator, such as tetryl, is needed to get all of the explosive involved.

Having a sensitive primary explosive next to your insensitive explosive defeats much of the reason for having an insensitive explosive in the first place. If the bomb or warhead was in a fire, it would still detonate. For this reason, initiators called "exploding bridgewire" detonators were developed and widely used in nuclear weapons. Exploding bridgewires are like blasting caps in that they are electrically detonated. The difference is that much more current is sent through them much more quickly. The wire itself literally explodes.

Luis Alvarez and Lawrence Johnson invented the exploding bridgewire detonator for the first plutonium bomb as part of the Manhattan Project. They needed much more precise timing than explosive primers could deliver, as the bomb depended on a perfectly spherical implosion to get the plutonium to critical mass.

In an exploding bridgewire detonator, a high voltage pulse vaporizes the wire in a few microseconds, and this sets off the secondary explosive directly, or a booster charge with a very high detonation velocity that then sets off the main charge. All of the bridgewires in a nuclear weapon (there were thirty-two in the first Fat Man tests) can be fired to within one hundred nanoseconds of one another. In an RDX explosive the detonation travels only 0.875 millimeters in that time, so the charges are set off at effectively the same time.

Newer versions of the detonator are exploding foil initiators, where a couple thousand volts of electricity from a capacitor explodes a foil bridge. These can directly detonate some of the new heat-resistant explosives like NONA.

One of the new less sensitive explosives being developed is 1,1-diamino-2,2-dinitroethane, known by the contraction DADNE or, more commonly, FOX-7.

FOX-7 is a much simpler molecule than many of the newer explosives (compare to CL-20, HMX, or TATB) but it is in the same class as TATB and shares its insensitivity. It is less dense than TATB (1.885 grams per cc, as opposed to 1.93) and has a lower melting point (238°C compared to 350°C), but it has a higher detonation velocity (8,870 meters per second, compared to TATB's 7,350 mps), approaching HMX (9,100 mps).

FOX-7 is insensitive to friction, and in drophammer tests, none of the samples detonated until the hammer was dropped from at least 63 centimeters. Larger crystals (250 microns instead of 70) were even more stable, at 79 centimeters. FOX-7 is also thermally 40% more stable than RDX and 60% more stable than HMX.

FOX-7 has been bonded with small amounts of the energetic polymer polyGLYN to make an insensitive PBX with a detonation velocity of 8,335 meters per second.

There are three new insensitive explosives used in the new composite mixture IMX-101

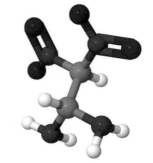

1,1-diamino-2,2-dinitroethane
(DADNE or FOX-7)

designed to be an insensitive explosive to replace TNT in army artillery shells. The IM stands for Insensitive Munitions.

The first explosive in IMX-101 is 2,4-dinitroanisole, known as DNAN.

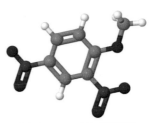

2,4-dinitroanisole (DNAN)

DNAN is not as powerful as TNT (its detonation velocity is 5,670 meters per second) and is less dense. As you might expect from its similarity to picric acid, it is also a decent dye. It apparently also makes a good insecticide. In mixes with other explosives, its castability makes it a good replacement for TNT. Molten DNAN dissolves RDX just as TNT does (not an issue in IMX-101, since it has no RDX, but useful in other formulas).

The second insensitive explosive in IMX-101 is 3-nitro-1,2,4-triazol-5-one, called NTO.

With a detonation velocity equal to RDX (which would make it about 8,750 meters per second), NTO is much less sensitive to heat, shock, fric-

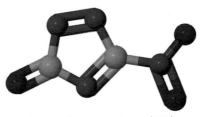

3-nitro-1,2,4-triazol-5-one (NTO)

tion, and impact. With a density of 1.93 grams per cc, it is denser than both RDX and HMX. It is made from inexpensive starting materials, and high yield processes are in place. With its high density and high detonation velocity, it makes up for some of the disadvantages of DNAN.

NTO was first synthesized in 1966 but the inventors did not investigate its use as an explosive. That task fell to Michael D. Coburn, who patented its use as an insensitive explosive in 1987. Working at Los Alamos National Laboratory in New Mexico, Coburn had been investigating several other insensitive explosives for use in nuclear weapons and deep drilling wellbores for geothermal energy and oil exploration.

The third ingredient in IMX-101 is nitroguanidine, known as NQ (even though there is no Q in the chemical name—the name NG, for nitroglycerin, was already taken).

Nitroguanidine is inexpensive (it is used as a fertilizer and in airbags) and is insensitive to heat, shock, and friction but has a high detonation velocity (8,200 meters per second). It has been used since the 1930s in triple-base (with nitroglycerin and nitrocellulose) smokeless powder, because it reduces the flame temperature and thus reduces the flash from the gun barrel (which can give away position at night). The lower temperature also reduces wear on large artillery bores. Like DNAN, it is also used as an insecticide, with effects similar to nic-

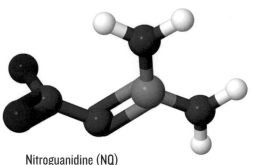

Nitroguanidine (NQ)

otine. Like the other ingredients in IMX-101, nitroguanidine has a low sensitivity to shock, heat, and friction.

With all three ingredients working together, the detonation velocity of IMX-101 comes out as the same as TNT, the explosive it was designed to replace. It is considered an EIDS—an Extremely Insensitive Detonating Substance. It is affordable and can be produced in TNT-like quantities. It can be melted and cast just like TNT, so no retooling is necessary.

IMX-101 is designed as an insensitive replacement for TNT. To replace the more powerful Composition B (60% RDX and 40% TNT, detonation velocity 7,980 meters per second), a more brisant explosive was needed. The result was IMX-104.

IMX-104 is a mix of DNAN, NTO, and RDX. It melts and casts just like Composition B and has a detonation velocity almost as good—7,400 meters per second.

The use of RDX in IMX-104 makes it more sensitive than IMX-101, but it is still much less sensitive than the Composition B it replaces.

A mix similar to IMX-104 is PAX-21 (RDX, DNAN, and ammonium perchlorate). The ammonium perchlorate was seen to be less environmentally friendly, so PAX-41 was developed. It is a mix of RDX, DNAN, and another insensitive explosive, N-methyl-4-nitroaniline, called MNA.

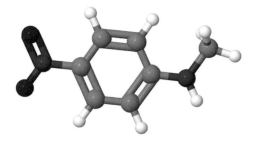

N-methyl-4-nitroaniline (MNA)

PAX-41 was designed as a replacement for Cyclotol, a more brisant mix of RDX and TNT than Composition B (Cyclotol has 65% to 85% RDX).

In 1901 the husband and wife team of Irma Goldberg and Fritz Ullmann described a new method for sticking organic molecules together using precipitated copper powder as a catalyst. They called it the Ullmann reaction, and named a similar reaction also using the copper catalyst the Goldberg reaction. Two molecules made up of a benzene ring with a halogen atom attached (chlorine, bromine, or iodine) can be joined together where the halogen atoms used to be.

The chemist Joseph Carl Dacons, working at the Naval Ordnance Lab in White Oak, Maryland, and the Naval Surface Weapons Center in Silver Spring, Maryland, used the Ullmann reaction to create several new explosive molecules that were heat resistant. Heat resistance is very important in missile warhead explosives, since the high speed of the missile generates very high temperatures as it travels through the air.

In 1960 Dacons and Mortimer Kamlet investigated the thermal decomposition of TNT in order to begin research on heat resistant explosives. Dacons published the eighth volume on heat resistant explosives, describing nonanitroterphenyl, or NONA.

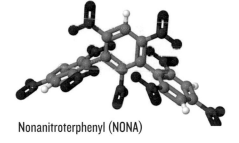

Nonanitroterphenyl (NONA)

You can see the three phenyl rings joined together and the nine nitro groups attached. This looks a little like three picric acid molecules stuck together. NONA was much more thermally stable than TATB, and had explosive performance and heat or friction sensitivity similar to Tetryl. In impact studies it was in between PETN and RDX in sensitivity.

NONA found use as the explosive to use in downhole well applications for the oil and gas industry, as the temperature miles underground is too high for conventional explosives.

In 1962 Dacons and J. M. Rosen published a study about how the structure of organic explosive molecules affects their thermal stability.

On August 31, 1963, Dacons filed a patent for an improved method for making NONA, using the Ullmann reaction.

On December 30, 1963, Dacons and two other chemists patented the process for making a heat-resistant explosive called DIPAM using older methods. On May 28, 1964, they filed a patent for a much simpler method, using the Ullmann reaction.

In the patent they describe how adding two amino groups to the existing explosive raised the melting point to where a thermally stable explosive resulted. The results of the research into the thermal degradation of TNT were starting to pay off.

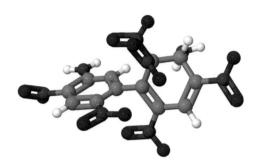

Diaminohexanitrobiphenyl (DIPAM)

DIPAM is related to another explosive molecule, hexanitrobiphenyl, or HNBP.

Hexanitrobiphenyl was one of the molecules synthesized by Fritz Ullmann using his eponymous reaction. He needed to use a solvent to slow the reaction of picryl chloride with his finely divided copper catalyst because

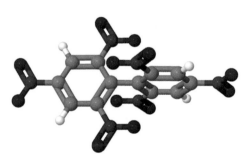

Hexanitrobiphenyl (HNBP)

otherwise the mixture would explode. Dacons reported that it was a heat insensitive explosive. He uses the acronym HNB, which unfortunately is also an acronym for several other explosives.

In 1965 Dacons and two other chemists reported on the improved method of making the heat-resistant explosive NONA using the Ullmann reaction.

On December 15, 1965, Dacons filed a patent for Dodecanitroquaterphenyl, where he had stuck together four phenyl rings, all nitrated. This was the heat-resistant explosive DODECA, which turned out to be cheaper to make than NONA, as it started with cheaper ingredients and had a high yield.

On January 18, 1966, Dacons and Kamlet reported creating PIPICL by joining picryl chloride with chloroanisole using the Ullmann reaction and then nitrating the result.

The two precursor molecules are shown below.

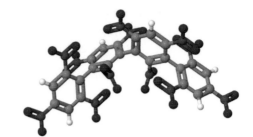

Dodecanitroquate`rphenyl (DODECA)

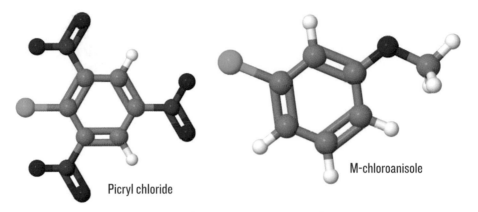

Picryl chloride

M-chloroanisole

The result looked a lot like a couple of picric acid molecules stuck together.

On November 22, 1967, Dacons filed a patent for the explosive octanitroterphenyl, or ONT. Looking for a more powerful heat-tolerant (yet still impact sensitive) explosive than NONA, Dacons again used the Ullmann reaction to make a very similar molecule, this time with one fewer nitro group. This decreased the impact sensitivity over NONA (hammer

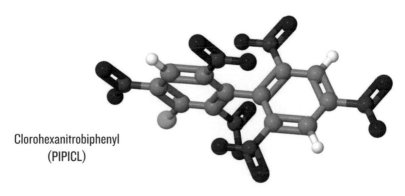

Clorohexanitrobiphenyl
(PIPICL)

impact detonation at 64 centimeters instead of NONA's 37 centimeters) and increased the heat tolerance over NONA by 20°C.

A year later, on November 20, 1968, Dacons filed a patent for tripicryltriazine, or TPT.

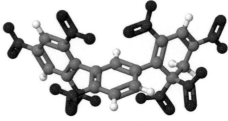

Octanitroterphenyl (ONT)

TPT is less impact sensitive (93 centimeter hammer drop) than ONT (64 centimeter) or NONA (37 centimeter), while being of comparable explosive force. While NONA is good to 300° Celsius and ONT to 320°C, TPT melts at 362°C.

Tripicryltriazine (TPT)

As a result of his many papers, books, and patents in organic chemistry and explosives, Joseph Carl Dacons is mentioned in several books and articles about African American inventors, and his patents are cited and referenced in much of the current literature on heat-resistant explosives.

The need for thermally stable explosives was recognized in many areas, and one of the better-funded labs was the Los Alamos National Laboratory (LANL), run by the University of California (although the lab is in New Mexico).

On May 17, 1967, chemist Michael D. Coburn filed a patent for a new explosive made at LANL called 3-picrylamino-1,2,4-triazole, nicknamed PATO. It was a heat-insensitive explosive but also impact insensitive, requiring a hammer drop from higher than 320 centimeters to cause it to detonate.

As PATO is made in a single-step process from inexpensive starting materials, it was expected to be a low-cost heat-resistant and impact-insensitive explosive. It has a high density of 1.94 grams

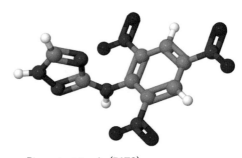

Picraminotriazole (PATO)

per cubic centimeter, melts at 310°C, and has a detonation velocity of 7,850 meters per second.

On March 26, 1971, Coburn filed a patent for another new explosive, called 2,6-bis(picrylamino)-3-5-dinitropyridin and given the shorter name PYX. The patent describes the unique characteristics of the new explosive, stating that PYX "is a moderately powerful explosive that is more thermally stable than any other explosive of comparable oxygen balance thus far reported. In addition to being chemically inert, the material has no vapor pressure at high temperatures. The compound is easily prepared in good yield from relatively inexpensive, commercially available starting materials."

PYX is stable to 350°C with an impact sensitivity of 63 centimeters in the hammer drop test and a detonation velocity of 7,448 meters per second.

Not only is it a good idea to have thermally stable explosives in intercontinental ballistic missile re-entry vehicles that get white-hot as they return to earth, but the LANL also has a Hot Dry Rock Geothermal Energy Development Program. Drilling through hot dry rock occasionally requires blowing up a stuck drill pipe

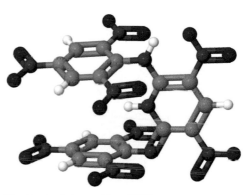

Picrylaminodinitropyridine (PYX)

in deep water-filled wellbores, where the water under pressure reaches temperatures above 320°C.

In 1979 Coburn was a visiting scientist at Elgin Air Force Base in Florida, where he started developing some new heat-resistant explosives. One of the explosives developed there was a mixture of 95% PYX and the polymer Kel-F 800. That polymer is basically Teflon with one of the fluorine atoms replaced by a chlorine atom. The air force was evaluating the mix as a high-temperature booster explosive, but the team back at Los Alamos tried it out, successfully, as a hot wellbore explosive.

Geothermal wells are not the only hot wells around. The deeper the drill goes, the hotter the rock gets—eventually you would get to the molten core of the earth. Explosives are used in all deep kinds of drilling operations, such as oil and gas exploration.

On September 1, 1976, Coburn filed a patent on the heat-resistant explosive ammonium 2,4,5-trinitroimidazole, ATNI for short. With an impact sensitivity of 50 centimeters and a detonation velocity of 8,560 meters per second, ATNI has the power of RDX but much better thermal stability.

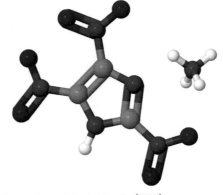

In the patent he describes potassium trinitroimidazole as the inspiration, a molecule with no great explosive potential because the potassium is "dead weight" and the rest of the molecule is

Ammonium trinitroimidazole (ATNI)

poorly oxygen balanced. Swapping the potassium for the ammonium ion fixes both of these problems in a single stroke.

In 1987 Coburn patented the use of NTO (shown earlier) as an explosive.

The most thermally insensitive explosive known at this time was N,N'-bis (1,2,4-triazol-3-yl)-4,4'-diamino-2,2',3,3',5,5',6,6'-octanitroazobenzene, known by the names DCONAB and BTDAONAB. It was synthesized by the Indian chemist Jai Prakash Agrawal, another pioneer in thermally stable explosives. Agrawal explained the four avenues to making an explosive less sensitive to heat, and it is useful at this point to list them each with examples.

One of the first heat-stable explosives was hexanitrostilbene, or HNS. On May 5, 1964, Kathryn G. Shipp of the Naval Ordnance Laboratory in White Oak, Maryland, filed a patent for the molecule, where her synthesis consisted of adding TNT to Clorox bleach. This explosive was vacuum tolerant and heat insensitive

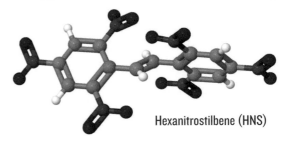

Hexanitrostilbene (HNS)

and was used as an explosive on the moon by the Apollo missions to perform lunar seismometry.

Previously, no one had been able to add more than five nitro groups to stilbene, since the nitro groups in place already prevent any more from being added. By starting with two molecules that already had three nitro groups each (trinitrotoluene) and then joining them, this problem was circumvented.

Hexanitrostilbene is an example of one of Agrawal's four avenues to heat stability: conjugation. The two TNT molecules have been joined together by a double bond.

A very *unstable* explosive is trinitrobenzene, or TNB.

Replacing each hydrogen atom with an amine group (NH_2) results in the very heat-stable molecule TATB (discussed earlier).

Trinitrobenzene (TNB)

This is an example of a second of Agrawal's four avenues to heat stability: adding amino groups. Adding amino groups to HNS gives us two of the avenues and a very heat-stable molecule, diaminohexanitrostilbene, with a much higher melting point.

A third of Agrawal's avenues is the condensation with triazole rings. This was used by Michael Coburn, who produced PATO that way (shown earlier, page 124).

The last of Agrawal's four avenues to heat stability is converting an acid to a salt. This sometimes backfires, as when picric acid is reacted with lead, copper, or iron. But reacting it with the ammonium ion to

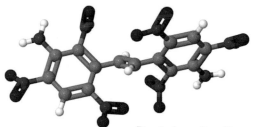

Diaminohexanitrostilbene

get ammonium picrate works well to heat stabilize the molecule.

In searching for a heat-insensitive replacement for PETN, where sensitivity to shock is an asset, making it useful in a detonator, USSR chemists (Sharmin, Buzykin, and Fassakhov) in 1965 synthesized dipicryloxadiazole, or DPO. With the thermal stability of HNS and NONA but the shock sensitivity of PETN, it was a direct hit. Now electric bridgewire detonators or exploding foil detonators could be made with a thermally stable booster explosive to set off the main charge of a heat-tolerant shock-insensitive explosive.

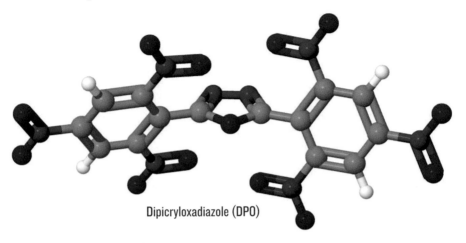

Dipicryloxadiazole (DPO)

The molecule is heat stable because it is conjugated—two nitrated benzene rings are joined. But the single bond between the two nitrogen atoms in the bridge between them is easily broken by shock, and once that one bond breaks, the whole molecule falls apart and an explosion ensues.

Increasing the density of an explosive is one way to make it more powerful. One way to increase the density is to make the molecule bend around and attach to itself in one or more places. Extreme examples of this are the cage-like molecules such as CL-20 (shown earlier, page 97), cubane, and adamantane.

Adamantane has a six-carbon base, from which three carbon atoms rise like pillars and then connect to yet another carbon to form a peaked roof.

Unlike many cage molecules, the bond angles in adamantane are not strained. This makes the molecule stable, as there is no energy waiting to be freed by a rearrangement of the molecule. It can be thought of as three six-carbon rings,

Adamantane

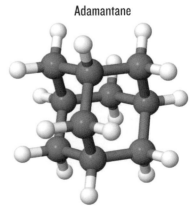

each ring sharing half of its carbon atoms with another ring. The name comes from the Greek word meaning diamond, as the arrangement of the atoms in space is the same as it is in diamond.

At $1,000 per kilogram, adamantane is not one of the cheapest starting ingredients for a modern explosive.

A molecule with a similar shape is hexamethylenetetramine, the base for RDX.

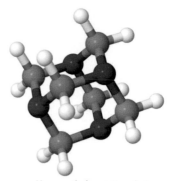

Hexamethylenetetramine

Taking advantage of the stability and density of adamantane, chemists Everett Gilbert and Gilbert Sollott filed a patent on October 14, 1980, for a way to add four nitro groups to adamantane to make an explosive out of it. The resulting molecule, tetranitroadamantane, is heat stable, not exploding until it reaches 400°C. In drophammer tests it stands up to a drop from 150 centimeters (TNT can withstand only 65 centimeters), making it impact insensitive as well as heat insensitive.

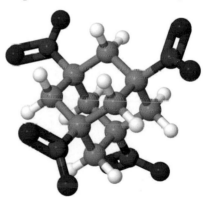

Tetranitroadamantane

SHAPED CHARGES

In 1792 the twenty-six-year-old German Catholic philosopher, theologist, and mining engineer Franz Xaver von Baader came up with a way of saving gunpowder in mines. The idea was to make a conical or mushroom-shaped hollow in the front of the gunpowder charge. His reasoning was not recorded, and it is not clear what effect it had, if any, but it was used in Norway and later in the mines of the Harz Mountains in Germany, but not since.

Gunpowder does not detonate by shock wave the way high explosives do, so the similarity in form of Baader's invention and later shaped charges may be coincidental, since in true shaped charges the effect is due to focusing the detonation wave, which gunpowder does not have.

The hollow in the gunpowder may well have been to allow the gases to cool a bit before hitting the rock, which, while reducing its effectiveness in moving rock, would have been important in reducing the chance of methane (called fire-damp in those days) explosions.

Things were a little different ninety-one years later, in 1883. The chief company engineer of the nitrocellulose factory of Wolff and Co. in Walsrode, Germany, was a man named Max von Förster. Von Förster tested a hypothesis "to give the detonating gases of gun-cotton a certain direction aiming toward the target" through a series of experiments with hollowed-out charges. Aiming at thick wrought-iron plates, he proved his hypothesis, saying, "Estimated on the whole, it appears that the effect of the hollow cartridge of the same size and less weight is superior to the full one of more weight."

Förster had discovered what is now called the *unlined cavity* principle. It is a shaped charge but only the explosive gases are propelled forward, and the effect is not nearly as powerful as the *lined cavity* shaped charge.

In 1885 Gustav Bloom patented a metal detonator cap with a hollow in it, to get a "concentration of the effect of the explosion in an axial direction." This was the first lined cavity shaped charge.

Three years later, in 1888, Professor Charles E. Munroe, a chemist working as a civilian for the US Naval Torpedo Station, found the same effect in a different way, apparently not having seen Förster's previous work.

In the February 1900 edition of *Popular Science Monthly*, Munroe describes his experiments:

> Among experiments made to demonstrate the resistance of structures to attack by a mob was one upon a safe twenty-nine inches cube, with walls four inches and three quarters thick, made up of plates of iron and steel, which were re-enforced on each edge so as to make it highly resisting, yet when a hollow charge of dynamite nine pounds and a half in weight and untamped was detonated on it a hole three inches in diameter was blown clear through the wall, though a solid cartridge of the same weight and of the same material produced no material effect.
>
> The hollow cartridge was made by tying the sticks of dynamite around a tin can, the open mouth of the latter being placed downward, and I was led to construct such hollow cartridge for use where a penetrating effect is desired by the following observations:
>
> In molding the gun cotton at the torpedo station, as stated above, a vertical hole was formed in each cylinder or block in which to insert the detonator, and in the final press a steel die was laid upon the cake so that an inscription in letters and figures was forced upon it. This inscription was indented in the cylinders and was raised upon the surfaces of the blocks. When the gun cotton was fired untamped, in testing it, the cylinder or block was usually placed with the inscribed face resting on a polished iron plate or iron disk, and after firing, if the gun cotton had detonated it was invariably found that not only was a vortexlike cavity produced below the detonator, but that the inscription on the gun cotton was reproduced on the iron plate, and, what was most singular, when the inscription was indented in the

gun cotton it was indented in the iron plate, and when the inscription was raised on the surface of the gun cotton it was reproduced raised on the surface of the iron plate.

In experimentally investigating this phenomenon I eventually soaked several cylinders in water, so that I could bore them without danger, and then bored holes of various diameters and depths in them, until in the last instance I bored a vertical hole an inch and three quarters in diameter completely through the cylinder. These wet cylinders were each placed on a similar iron plate, a similar dry disk was placed on each as a primer, and they were successively fired, when it was found that the deeper and wider the hole in the gun cotton the deeper and wider were the holes produced in the iron plate, until when the completely perfo-rated gun-cotton cylinder, from which at least half of the weight of explosive had been removed by the bor-ing, was fired, the iron plate was found to be completely perforated.

Advantage was taken of this action of the rapidly moving molecules to produce some beautiful effects by interposing laces, coins, leaves from the trees, and stencils with various devices cut in them between the base of the gun cotton and the iron plate, for after the detonation of the gun cotton the objects were found to be reproduced upon the iron with the utmost fidelity and in their most delicate parts, and the impressions were raised upon the iron as the objects had been before the explosion.

In the first experiment the tin can acted as a liner. A liner changes the dynamics of the shaped charge enormously. In the later experiments the charges were unlined, and the crucial distinction seems not to have registered with Munroe. In a lined cavity shaped charge, the liner is crushed and reshaped into a projectile jet of metal, traveling at a respectable multiple of the detonation speed of the explosive. In some modern shaped charges a detonation velocity of less than 10 kilometers per second can produce jet velocities in the range of 90 kilometers per second.

On December 11, 1911, a UK patent was issued to the company WASAG, of Berlin, Germany, for a cavity shaped charge, not mentioning any prior work on the subject. It had versions with paper linings, linen linings, and metal linings, but stated that these were merely for waterproofing, and metal linings would be used in shells fired by cannons to prevent the lining from deforming due to the sudden acceleration. The patent was tried in torpedo tests in 1913 and shown to be effective but the British decided not to use it, and World War I had no shaped charges.

On June 25, 1936, an Englishman, Professor R. W. Wood of Johns Hopkins University, published a paper in the *Proceedings of the Royal Society* (as a foreign member of the Society) titled "Optical and Physical Effects of High Explosives."

In that paper, he describes how he came to study "the plastic flow of metals" in explosives, due to an unfortunate accident that happened to a young woman in England:

My interest in the study of the effects produced by high explosives originated in the investigation of "evidence" in a number of murders by bomb, and more especially in connexion with a most unfortunate and unusual accident which resulted in the death of a young woman who, on opening the door of the house furnace to see if the fire was burning properly, was struck by a small particle of metal which flew out of the fire and penetrated the breast bone, slitting a large artery and causing death in 2 or 3 minutes from internal haemorrhage. The particle, which was not much larger than a pin-head, was submitted to me for identification, and though its form resembled nothing with which I was familiar, I surmised that it was probably a part of a dynamite-cap or "detonator" used for exploding the dynamite charges in the mines, which, by some carelessness on the part of a miner, had been delivered intact with the coal.

These detonators are spun from very thin sheet copper and consist of a tube about 5 mm. in diameter and 40 mm. in length. The head is formed into a shallow cup, as shown in cross-section in fig. 1, and the tube is charged with mercury fulminate and fired by an electrically heated wire. It seemed probable that the solid pellet of copper, recovered during the autopsy, had been formed in some way from the concave head of the detonator by the enormous instantaneous pressure developed by the detonation of the fulminate.

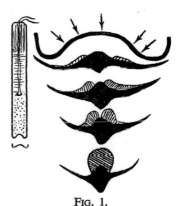

Fig. 1.

I accordingly suspended one about 2 feet above a large earthenware jar holding about five gallons of water, pointing the head downwards. On firing the detonator the jar was shattered into a dozen or more pieces by the pressure wave exerted in the water by the passage of the small copper fragment (the head of the detonator) entering into the water with three times the velocity of a rifle bullet, just as a milk can filled with water is burst open when the bullet of a high powered rifle is fired through it. The minute fragment of copper which was found in the ruins of the jar matched perfectly the fragment found during the autopsy but bore no resemblance to the original head of the detonator which is about the size of, and resembles closely the cap of, a shot-gun shell after it has been indented by the firing pin. It was a pear-shaped pellet of copper, surrounded at the middle by a skirt of thin copper of a diameter considerably less than that of the original head of the detonator. This looked interesting, and a study of exploding detonators was commenced with a view of finding out how the forces operate to mould this solid pellet.

Wood had discovered what is now referred to as *explosively formed projectiles*. But no military seemed to understand the importance of the effect he described.

That changed during World War II, when Dr. Franz Rudolph Thomanek, on February 4, 1938, discovered the importance of the metal lining. It seems that everyone involved was rediscovering the effects without knowledge of prior work (although Wood, after discovering the effect, did research and found Munroe's documentation of it). Between then and 1943, Thomanek experimented with different hollow shapes and different liner thicknesses, including tapering the thickness.

In Zurich in 1937 a Swiss engineer, Dr. Henry Hans Mohaupt, also came up with a metal-lined hollow cavity shaped charge, and tried to get the British government to pay him for the expense of developing it. They refused to pay, as hollow explosives and their effects were well known by then, not seeing that the metal liner was an important new invention. But back home in Britain they quickly discovered that metal-lined cavities made a huge difference.

Mohaupt brought his idea to the United States on October 18, 1940. First developed as a rifle grenade, the idea morphed into a rocket-propelled munition with a shaped charge armor-piercing warhead that became known as the "bazooka," named after a pipe-like musical instrument in a then-current comic book. Despite being a morale-boosting weapon loved by the public, the bazooka did not actually work well as a shaped charge device, as the fuse was too slow and the charge only went off after the warhead had slammed into the side of the tank and deformed. Some spalling inside the tank occurred but no penetrating hole in the armor came to be.

The US Navy had better luck with the shaped charges used in torpedoes and antisubmarine bombs.

In the Korean War this was remedied with the 3.5-inch Super-Bazooka. The US Navy at this time was having trouble with its 5-inch HVAR rockets just bouncing off the armor of the Soviet T-34 tanks used by the North Korean Army. The job of fixing this fell to the explosives experts at the navy's China Lake base. In twenty days they developed new warhead designs from scratch, incorporating shaped charges, for the 6.5-inch anti-tank aircraft rocket (ATAR).

Nonmilitary uses of shaped charges began to find use during the 1950s (actually starting as early as 1948). Oil and gas exploration companies had been using well casings since the 1920s to prevent rock around the borehole from collapsing inward and blocking the bore. This created a problem since now the oil or gas had no way to enter the casing. Casing perforators were invented where bullets were inserted into carriers and lowered into the well, then set off so the bullets would pierce holes in the casings and allow the oil or gas to flow through the holes.

Bullets created problems of their own, however, as they would often end up blocking the holes they had just created. Enter shaped charges. The metal liner of the shaped charge could be designed to disintegrate into tiny pellets once it had penetrated the casing and the rock behind it, or it could be made of a material such as aluminum, which would vaporize and burn away after penetration. This process is called perforating.

An array of shaped charge explosives arranged in a cylinder with the charges spiraling around the outside forms the device that is lowered into the well. Early devices use sensitive primary explosives in the detonators, triggered by heating a wire to incandescence with a small battery. These proved susceptible to detonation by stray radio traffic, as the mile-long wires needed to reach the bottom of the well acted as long antennas. Strict radio silence had to be enforced to prevent accidental ignition.

Later devices used exploding foil detonators that require some three megawatts of power to initiate (delivered by high-voltage capacitors in the device). This is millions of times the amount of power that can be delivered by communications radios, and now radios and cell phones can be used without fear of causing accidents.

The metal liners in the shaped charges form projectile jets of explosively formed metal that penetrate through the casing and far into the surrounding rock. The length of the jet is a function of the diameter of the metal liner. Jet lengths are generally many times the charge diameter and carry enough kinetic energy to penetrate even farther.

In a shaped charge the shape of the metal lining has a big effect on the result. A slightly domed disc, like that in Professor Wood's fulminate detonator, is reduced in diameter and increased in thickness to produce an *explosively-formed projectile*, or EFP. If the liner is instead a cone, it produces a thin jet of metal that can be very long. These jets are very good at piercing holes in thick metal or rock at close range. EFPs are better when the range is greater (the term of art for this is a larger *standoff*).

Some military designs get quite complicated. The shape of the detonating wave is carefully controlled by changes to the lining and by using wave-shaping devices. A wave-shaping device changes the detonation wave, usually to increase the forward velocity of the tip of the jet. In a simple jet-forming shaped charge, one might see a copper cone with an angle of 42 degrees at the forward end of the high explosive charge. In the drawing the white area is open air, the gold color is the explosive, and the orange is the copper. The copper might be 2.25 millimeters thick, and the diameter of the charge 85 millimeters.

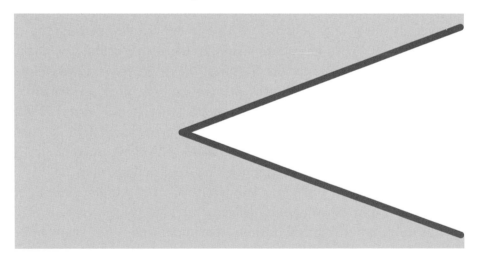

The jet velocity of such a charge might be 8,000 meters per second, about the same as the speed of the detonation wave itself. The maximum pressure on the copper liner is about 12 million pounds per square inch.

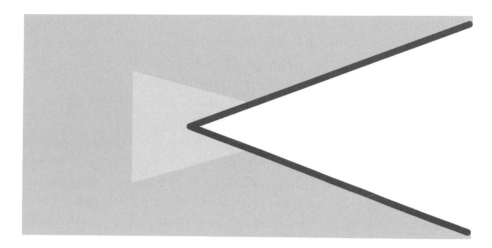

If a wave-shaper (in this case, just an air cavity, but other light inert fillers like foam can be used) is placed behind the apex of the cone, the maximum pressures on the cone go up to 27 million pounds per square inch, and the tip of the jet increases to 10,000 meters per second. In addition, there are particles of copper that precede the jet, due to their velocity of 14,000 meters per second.

The maximum pressure occurs twelve microseconds after detonation. The maximum velocity of 14 kilometers per second is achieved in thirty-two microseconds after detonation. That comes out to an acceleration of 437.5 million meters per second squared, or over 44 million times the force of gravity.

The copper at this point has not vaporized or even melted. It has been transformed into a metal spear with enormous kinetic energy. When it encounters tank armor, or the wall of an oil well casing, it transfers its momentum to the material, which also deforms hydrodynamically (like a liquid), making a hole. Both the jet and the target react as liquids at these speeds.

The length of the jet may be several times the diameter of the charge. The tail of the jet is moving at a much slower speed, perhaps half as fast as the tip. In oil drilling applications the jet may be designed to break up into small particles that do not clog the hole in the casing.

The liner can be made of a number of different materials depending on the desired outcome. Very ductile metals such as copper (or better yet, a composite material of tungsten fibers in an amorphous metal alloy designed to be extremely ductile) can form long jets before breaking up into smaller particles that more quickly lose their velocity in air. The denser the lining material is compared to the density of the object being penetrated, the better the penetration. This is why tungsten fibers are sometimes added to the more ductile metals.

The longer range explosively formed projectiles are often made of copper formed into a hemisphere, just like the one that started Professor Wood's investigations (but larger, of course). Instead of a range of less than a meter, as jets provide, EFPs can travel 100 meters or more. Using special wave-shapers, the projectile can be formed precisely, even to the point of having stabilizing fins and rotation.

Besides opening up well casings and destroying enemy tanks, shaped charges are used in the military to safely defuse other explosives. The basic copper cone jet-producing shaped charge creates a very thin jet that moves so fast that the target is just pierced, not moved. Thus, the jet can be aimed at the fuse mechanism of a bomb and miss the primer and secondary explosive.

Making the metal liner out of magnesium makes the jet less penetrative (magnesium is less dense and less ductile than copper). Magnesium also is much more reactive than copper, burning in air and stealing oxygen from most explosive molecules. A magnesium shaped charge jet can cause the TNT or RDX in the bomb to burn without causing a detonation. Sometimes aluminum is used, for similar reasons. It is less reactive but more ductile and cheaper.

EFPs allow the bomb or mine to be blown up from a distance since they have better standoff than shaped charge jets do. De-arming a mine or an improvised explosive device from a meter or two away can sometimes be done without setting the bomb off.

Replacing the metal lining with one made of water (sandwiched between two conical plastic liners, so the shaped charge effect still works) makes a device that can still penetrate a half-inch of steel, but then scatters the explosive without detonating it.

To counter shaped charge weapons, tank armor has improved. The use of composites instead of steel, the use of layers of armor separated instead of a single thickness, and the use of so-called "active" armor made of explosive material that disrupts the jet or destroys the EFP, have all been added. In turn, makers of antitank weapons have developed ways to counter these innovations.

Tandem-shaped charge weapons, consisting of two warheads stacked one in front of the other, allow the forward one to travel faster, setting off the active armor or destroying the first layer of composite armor. The slower moving warhead comes in right behind it, and finds a hole where the first layer of armor used to be.

Staging the timing of the explosion in a shaped charge can make the jet fragment, so that the fast-moving tip destroys one layer of armor and the slower-moving tail then finds a hole.

Specialized shaped charges include linear versions where instead of a cone, a long V-shaped liner is backed by an equally long explosive charge. These can cut along a line instead of piercing a hole, and can be used to cut large sheets of metal into smaller pieces. Special rebar cutting charges, using the same principle, can be clamped onto inch-thick or larger rebar, cutting them apart. These can also be used to cut thick steel cables.

Another special case of shaped charge is the explosive lens, used in nuclear weapons to implode the nuclear material and cause a chain reaction. The cone is made up of two different explosives instead of an explosive and a metal liner. The explosives have different detonation velocities. The shape of the interface between the two explosives acts like the curved surface of an optical lens, bending (refracting) the wave front of the explosive detonation in the same way glass refracts light.

To make a spherical wave front, so the nuclear core is imploded evenly, the shape of the interface is a hyperboloid. If a flat wave front is needed, the interface would take the shape of a parabola. Just like in optical lenses, multiple detonation velocities (from multiple explosive mixes) can fine-tune the wave front (in the way achromatic lenses do for light).

In a nuclear weapon a pair of precisely shaped wave fronts from two explosive lenses takes the place of thirty-two or ninety-six separate single charges and still provides an improved spherical implosion.

TYPES OF EXPLOSIVES

So far this book has discussed explosives in the order in which they were invented, and later, to some extent, by their sensitivity to heat, friction, or shock. In this chapter I will discuss the classes of explosive, which depend on the types of molecules used (their chemistry) and the mix of different molecules used. I will be primarily discussing high explosives since almost all explosives since black powder are of this type.

NITRATES

The nitrate-based explosives contain the nitrate group, NO_3. The simplest is ammonium nitrate.

Ammonium nitrate is a salt. A salt is what you get when you combine an acid (in this case nitric acid) with a base (ammonia). In a salt the ion from the acid is negatively charged, and the ion from the base is positively charged. In a water solution the water molecules (which have a positive end and a negative end) surround the ions and mask their

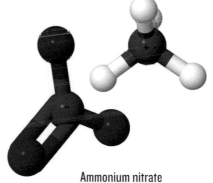

Ammonium nitrate

charges from one another, and they lose their mutual attraction and dissolve. Take away the water and the ions arrange themselves in a regular order (thus falling into a lower energy state) and crystallize.

Ammonium nitrate is explosive, unlike metal nitrates such as potassium nitrate (saltpeter). This is because the ammonium ion provides fuel in the form of those four white hydrogen atoms you can see in the image.

When detonated, the reaction products are all gases, such as water vapor, nitrogen gas, and oxides of nitrogen. There is no carbon, so there is no black sooty smoke.

Ammonium nitrate was not used as an explosive before the invention of a way of combining hydrogen with the nitrogen in the air to form ammonia gas. The first process to do that was invented by the German chemists Adolph Frank and Nikodem Caro in 1895. Their process used calcium carbide to react with nitrogen gas at a temperature of 1,000°C. The end product was calcium cyanamide, and the Frank-Caro process is often known as the cyanamide process.

Calcium cyanamide

A second process, the Haber-Bosch process, was invented later and was more economical. Lesser-known processes like the arc process (simulating how lightning makes nitric acid), the nitride process (reacting steam with nitrides), and the Bucher cyanide process (nitrogen, coke, and sodium carbonate react to make sodium cyanide, which steam converts to ammonia) were more costly and disfavored.

Before these processes were invented, nitrates were mined from dried bird guano deposits in deserts or leeched from compost heaps.

Nitrogen in the air is very stable. The two nitrogen atoms in N_2 are connected by a very strong triple bond, and it takes a lot of energy to tear that bond apart. In nature this energy comes in the form of lightning. The intense electrical discharge splits the nitrogen molecules, and some of them combine with the second-most plentiful component in air, oxygen. This forms nitric acid, which is quickly diluted by the large amounts of moisture in the air and comes down as very weakly acidified rain. Plants take up the nitrate ion as fertilizer and use it to make proteins and other biological molecules.

Nitrates are so important as fertilizer and oxidizer in black powder that the demand began to outstrip the supply in the 1800s. Chemists began looking for ways to "fix" (make available in other forms) the plentiful nitrogen gas in the air.

In the summer of 1909 two German chemists, Fritz Haber and Robert Le Rossignol, demonstrated their way to do this, using high pressure (around 200 atmospheres) and an iron catalyst promoted by potassium hydroxide. Their laboratory-scale device took an hour to make half a cup of liquid ammonia gas (under pressure to liquefy it). Only about 15% of the original gases combine, so the rest of the gases are cycled through the process repeatedly until the conversion rate gets to about 98%.

As you would expect from looking at the ammonium ion (or the formula NH_4), it takes four times as much hydrogen gas (by volume) as nitrogen gas to make the resulting ammonia. The reaction is *exothermic*, meaning it creates heat. Removing the heat allows more of the reaction to go to completion and fewer of the ammonia molecules to break apart into nitrogen and hydrogen gases. However, cooling the gases makes the reaction go too slowly. The compromise is to keep the temperature at somewhere between 400°C and 450°C, which is why only 15% of the reaction completes. Cooling the gases (still under high pressure) then allows the ammonia to be removed, and the remaining gases can once again be heated, producing 15% of ammonia again.

The number of molecules of ammonia is half the number of molecules in the original gases. This means the pressure will drop by half. Dropping pressure favors breaking apart the ammonia to make the original gases again. The answer to this problem is to raise the pressure to 200 atmospheres or so to move the equilibrium toward forming more ammonia. The higher pressure increases the rate that the molecules react, since they encounter one another more frequently.

High pressures are costly to produce, and require expensive equipment that can handle the pressure. The reaction would go faster at higher pressure, but 200 atmospheres was the sweet spot economically.

As the reaction product is cooled, the pressure would normally drop. Keeping the pressure high allows the ammonia to be liquefied at room temperature and separated from the unreacted gases, so they can be put through the cycle again.

Haber sold the process to the German chemical company BASF, who put Carl Bosch in charge of scaling the process up to meet industrial demand. He succeeded the following year. Both Haber and Bosch ended up getting Nobel prizes for their work on the Haber-Bosch process.

During World War I the Allies cut Germany off from the cheap Chilean nitrate sources. The German war effort depended on the Haber-Bosch process to make the explosives needed.

Ammonia can be oxidized to form nitric acid. That acid can be used to make explosives and fertilizer. And when the acid reacts with yet more ammonia, the result is ammonium nitrate, both a fertilizer *and* an explosive.

Ammonium nitrate is seldom used by itself as an explosive. It has a positive oxygen balance, so adding a fuel makes it a more powerful explosive. Ammonium nitrate mixed with fuel oil (94.3%/5.7%) is a cheap and common high explosive used mostly in mining. The mixture has a low detonation velocity (3,200 meters per second) but produces a lot of gas, giving it what explosives experts call "heave," the ability to move lots of rock. The mixture is known as ANFO and makes up an estimated 80% of all the explosives use in the United States. It is not easily detonated, making it what is known as a tertiary explosive, one that is normally set off by a secondary explosive (which is in turn set off by a primary explosive). Adding finely powdered aluminum makes it detonate more easily and can increase the power by up to 30%. Using nitromethane instead of fuel oil is another way to make the mixture more powerful. That mixture is known as ANNM.

Of course, ammonium nitrate is only the simplest of the nitrate explosives. Cellulose nitrate (guncotton) and glyceryl trinitrate (nitroglycerin) have already been discussed, as have PETN and polyGLYN.

Another simple explosive nitrate is methylammonium nitrate. The extra carbon and hydrogen atoms give it a better oxygen balance than ammonium nitrate, but it is not as cheap. It was used as an explosive by the Germans in World War II, and is now used only in water gel explosives such as Tovex Extra, which have largely replaced dynamite as commercial explosives. The "Extra" is methylammonium nitrate, replacing the more expensive TNT used in the original Tovex. The result is not just cheaper, it has a better oxy-

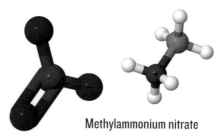

Methylammonium nitrate

gen balance, and can be detonated by blasting caps even in narrow drill holes (like the dynamite it replaced).

Water gel explosives are emulsions or slurries of nitrates in water, and they can be used in wet mining situations where other explosives required expensive drying before use. Various ingredients besides the nitrates are added (fuels like nitromethane, powdered coal, aluminum powder, thickening agents, etc.).

Urea nitrate is another simple explosive nitrate. Used as a cheap fertilizer, it is widely available and cheap, making it a favorite of amateur explosive enthusiasts and makers of improvised explosive devices (IEDs) in Pakistan, Iraq, and Afghanistan.

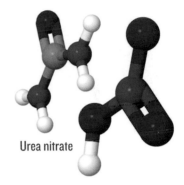

Urea nitrate

Trimethylolethane trinitrate is similar to nitroglycerin. It is sometimes used in propellants (smokeless powders) instead of nitroglycerin because it erodes gun bar-rels less and produces less muzzle flash. However, like nitroglycerin, it is being replaced by less sensitive, safer explosives.

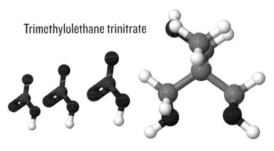

Trimethylolethane trinitrate

Erythritol tetranitrate is similar to the other tetranitrate discussed earlier, PETN. It has a high detonation velocity (over 8,000 meters per second) but is more sensitive to friction and impact than PETN. It is not used commercially or militarily, but it is a favorite of amateurs since it is easy to make and the starting point, erythritol, is a natu-

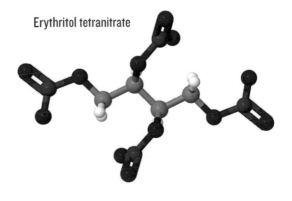

Erythritol tetranitrate

ral sweetener and easy to find. It has a long shelf life, like PETN. Unlike PETN, it has a positive oxygen balance, meaning extra fuel, such as aluminum powder, or an explosive with a negative oxygen balance, like TNT or PETN, could be added to improve the performance.

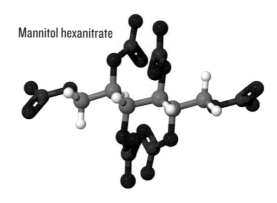

Mannitol hexanitrate

Mannitol hexanitrate is a nitrated sugar. It is a powerful secondary explosive (detonation velocity 8,260 meters per second, 1.7 times more powerful than TNT). It is more stable than nitroglycerine and used as the powder in detonators.

An interesting nitrate is guanidinium nitrate. As a propellant in rockets, it produces a lot of gas at relatively cool temperatures. It is used in toys as the propellant in Jetex engines.

It is also known (incorrectly, but widely) as guanidine nitrate. Completing the circle, it can be made by reacting calcium cyanamide with ammonium nitrate.

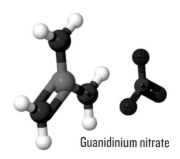

Guanidinium nitrate

NITRIDES

Another class of explosives is the nitrides. The "fulminating silver" discussed earlier is silver nitride, Ag_3N. It explodes violently, leaving metallic silver and nitrogen gas. Silver nitride is formed when an excess of aqueous ammonia (NH_4OH) is added to silver nitrate and allowed to stand (the reaction goes through stages, first producing silver oxide, which reacts with the extra ammonia to form diamine silver hydroxide, which gradually decomposes to form silver nitride).

Another very sensitive explosive nitride is mercury(II) nitride. Made by adding mercury oxide to aqueous ammonia at 10°C and then heating, it also explodes quite violently from the heat of just a slight touch. Nitrides are often too sensitive to be useful explosives.

A third class of explosives is the fulminates. I have discussed mercury fulminate, which found wide use as a primary explosive in guns. (It is now illegal in most countries.) Platinum fulminate is also explosive and, as you can imagine, a little too expensive to find wide use. It was discovered by Edmund Davy, the cousin of Sir Humphry Davy, the person who damaged his eyes while experimenting with explosive nitrogen chloride. Edmund Davy also discovered acetylene, and started the practice of bolting zinc onto ships' bronze rudders to prevent corrosion.

Potassium fulminate is another primary explosive that found use in early percussion caps. It is less sensitive than mercury fulminate. Silver fulminate is so sensitive it will go off under its own weight if too much is stored at a time. It is used in firework toys called bang-snaps, where a little of it (80 micrograms) is mixed with sand and wrapped in tissue paper. On hitting the ground, or being rubbed between the fingers, a loud snap occurs.

The acetylene that Edmund Davy discovered acts as a weak acid. It will combine with some metals to make explosive acetylides. I have already discussed copper acetylide and silver acetylide. Silver acetylide is formed by bubbling acetylene gas through a solution of silver nitrate. That process was discovered in 1866 by the famous French chemist Marcellin Berthelot.

NITRO

The nitro group (NO_2) forms a large group of famous explosive compounds. I have discussed TNT, TATB, TNP, nitroguanidine, and others. Some other nitro compounds are nitromethane and nitroethane, used in racing fuels, or mixed with ammonium nitrate to form explosives.

Note the difference between the nitro group and the nitrate group (the latter

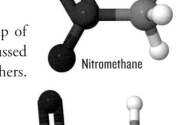

Nitromethane

Nitroethane

has an extra oxygen). Many nitrate explosives have common names that make them sound like they are actually nitro explosives (nitrocellulose, nitroglycerin, nitrostarch, etc.). One can look at the structural formula to quickly tell the difference.

Another nitro explosive is nitrourea.

Nitrourea

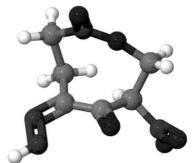

While nitrourea is a respectable explosive, it degrades quickly in contact with moisture and is not used commercially as an explosive.

More than one nitro group can be added to a molecule. One of the early explosives having two nitro groups is diazodinitrophenol (DDNP, or Dinol). It was discovered in 1858 by Peter Griess, (who later discovered lead styphnate, another primary explosive). His discovery of the diazonium salts (of which DDNP is one) made him famous in chemistry circles and got him a job in England at the Royal College of Chemistry. Diazonium salts are widely used in making dyes.

DDNP is a less sensitive primary explosive than mercury fulminate, and is thus somewhat safer to use. It is also more powerful, allowing it to set off the new less sensitive secondary explosives more easily. When unconfined, it simply burns with a quick flash like guncotton. If confined, as in a firecracker, it can be set off with a burning fuse. Since DDNP contains no lead, it is less toxic than lead azide or lead styphnate, and has been substituted for those in primers and detonating caps.

Diazodinitrophenol (DDNP)

Adding three nitro groups to a molecule is more difficult because the nitro groups do not respond well to crowding. However, the class of explosives containing three nitro groups is one of

Lead styphnate

the largest and includes TNT, RDX, picric acid, Dunnite, and Tetryl. Another notable trinitro explosive is lead styphnate, mentioned several times earlier as a common primary explosive.

One very interesting nitro group explosive is octanitrocubane, or ONC. The molecule cubane is fascinating all by itself. It is a cube of carbon atoms, each one bonding to another carbon with three bonds and to a hydrogen atom with the remaining bond. It was first synthesized by Philip Eaton of the University of Chicago in 1964 in a remarkable fifteen-step sequence.

Cubane is sometimes called the "impossible" molecule since the 90-degree bond angles of the eight carbons are extremely strained. As previously noted, the energy stored in such strained bonds helps molecules to rearrange, releasing that energy to add to the explosive force. Bond angles between three carbon atoms are most relaxed at 109.5°, almost 20° more relaxed than the tightly wound cubane is.

Since the original fifteen-step synthesis, several improved methods with fewer steps have been developed. Cubane is now being produced in multi-kilogram quantities by Fluorochem in California but is still quite expensive.

Despite cubane's strained bonds, it is highly stable. This is because there are no kinetically viable paths by which cubane can rearrange thermally. Just breaking one bond makes only a small change in the structure, and the result is still quite strained. It takes over 230°C before cubane decomposes, and even then, it does so slowly.

Cubane

Eaton did not stop there. Because cubane is very dense for a hydrocarbon (1.979 grams per cubic centimeter, so it sinks in water), and it has highly strained bonds, the US Army Armament Research and Development Center came to Eaton with a proposal. They wanted him to create an explosive out of the molecule by adding eight nitro groups to it, one at each corner of the cube.

The resulting molecule would have some interesting properties. Since there is no hydrogen, it would explode to create eight carbon dioxide molecules and four nitrogen gas molecules, but no water. Thus, there would be no smoke or steam produced, making it an excellent propellant for a rocket (no telltale plume behind it, making it difficult to track). The resulting gas would expand to 1,150 times the volume of the original explosive. The energy release would be phenomenal, and it would have a perfect oxygen balance, with no remaining fuel left over after the explosion and no remaining oxygen.

In addition to being very powerful, octanitrocubane would be an insensitive explosive, so safe to handle one could hit it with a hammer and it would not explode, and light it with a match and it would just burn without detonating. It was predicted to be less shock sensitive than CL-20.

When a nitro group is attached to a carbon (as it is in TNT), the molecule is more stable than when it is attached by way of an oxygen (as it is in nitroglycerin), or when it is attached by way of a nitrogen (as it is in RDX, HMX, and CL-20).

The density of octanitrocubane helps to increase the detonation velocity. So does the high weight of the product gases (CO_2 weighs 44 atomic mass units, and N_2 weighs 28, while H_2O only weighs 18). The high number

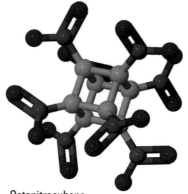

Octanitrocubane

of reactant gas molecules is a third factor. The high amount of energy released is the fourth part of the equation for calculating the detonation velocity, so octanitrocubane wins on all counts.

The detonation pressure is also important in an explosive, and it goes up with the square of the density. Octanitrocubane wins again. No other combination of carbon, nitrogen, and oxygen has a density as high as octanitrocubane. Compared to TNT, octanitrocubane is 2.38 times more powerful.

There is at least one more powerful explosive, however. The step just before octanitrocubane produces heptanitrocubane. It has one hydrogen atom. It is not as oxygen balanced as octanitrocubane. It is not so beautifully symmetrical. But it has one advantage that tips the balance. It is a little bit denser (2.028 grams per cubic centimeter). And, because the detonation pressure is a factor of the density squared, that makes enough of a difference to make heptanitrocubane the winner. This may not last long, however, as theoretical calculations show that there should be a crystal form of octanitrocubane that has a density of 2.1 grams per cubic centimeter. That crystal structure is different from the one currently being tested, and crystallizing the compound in some other way (different solvents, temperatures, or under pressure) may be the key to getting the higher density. The density of the densest form of CL-20, by comparison, is 2.044, which is denser still (but it is not oxygen balanced, as it has 11% too little oxygen).

Octanitrocubane is currently fabulously expensive to make. One of the cheapest cubane compounds is sold for $40,000 a kilogram, and octanitrocubane is harder to make than that. But theoretical calculations show that there might be a direct synthesis path to octanitrocubane by combining four dinitroacetylene molecules. This synthesis releases energy, so it should be chemically favorable. There is one catch—at the time of this writing, no one has yet made any dinitroacetylene.

NITRAMINES

A subclass of nitro group explosives is the nitramines. Nitramines (or nitroamines) are a nitro group connected to a nitrogen atom. The simplest member of the group is nitramide, where the extra nitrogen has two hydrogen atoms to complete the molecule.

Nitramide

Several of the most powerful and prominent military explosives are nitramines, such as RDX, HMX, Tetryl, nitroguanidine, and CL-20.

Some of the new nitramine explosives combine the stabilizing effect of pyridine

Nitroguanidine

(the central ring in the explosive PYX) with nitra-mines to get an insensitive explosive with the power of HMX.

One of those new explosives is diaminodiamino-trinitropyridine, or DADNPO.

On October 10, 1989, Thomas G. Archibald, working at explosives maker Flu-orochem, Richard Gilardi (Naval Research Laboratory), K. Baum (Fluorochem), and Clifford George (NRL) described a new nit-ramine explosive in the *Journal of Organic Chemistry*. Perhaps the simplest example of a nitramine, the compound, called trini-troazetidine, or TNAZ, is a high–energy density molecule with a high oxygen bal-ance. (Gilardi worked with Philip Eaton on octanitrocubane.)

On June 7, 1993, Michael A. Hiskey and Michael D. Coburn filed a patent for making TNAZ from inexpensive ingredi-ents with better yields. Not to be outdone, Archibald and three other chemists filed a patent on May 15, 1995, for a process that has high yields and uses "mild conditions."

Pyridine

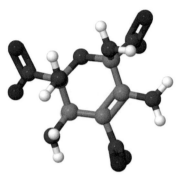
Diaminodiaminotrinitropyridine (DADNPO)

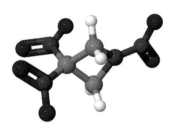
Trinitroazetidine (TNAZ)

TNAZ has nearly the energy density (96%) of HMX and melts without decomposing at the temperature of boiling water, making it melt-castable, like TNT (but with half-again as much power). Because it has a high oxygen balance and the reaction gases are very hot, it is expected to do well in composite explosives. While its shock and friction susceptibilities are similar to other nitramines, TNAZ is significantly less heat sensitive than RDX, HMX, or CL-20. No deto-nation occurred at temperatures below 199°C.

As a molecule with strained bonds caused by the four-atom ring, TNAZ has extra stored energy that is released when it detonates or burns, like CL-20 and octanitrocubane. Other selling points for TNAZ as an explosive or as a propellant are that it does not attract moisture and is compatible with aluminum, brass, steel, and glass, unlike more chemically reactive explosives.

While the nitramines are generally very high energy density molecules, the record holder is another nitro group explosive, hexanitrobenzene, or HNB.

Hexanitrobenzene (HNB)

Hexanitrobenzene was first formulated in Germany during World War II, but it is too sensitive to light to be used in commercial or military explosives.

The following table shows some of the higher density explosives and compares them to HMX:

HNB	1.92 grams/cc	1650 calories/ gram	3251 calories/cc	115% of HMX
CL-20	1.96 grams/cc	1500 calories/ gram	3060 calories/cc	109% of HMX
HMX	1.89 grams/cc	1480 calories/ gram	2816 calories/cc	100% of HMX
TNAZ	1.83 grams/cc	1465 calories/ grain	2710 calories/cc	96% of HMX
PETN	1.73 grams/cc	1490 calories/ gram	2652 calories/cc	94% of HMX
TNT	1.53 grams/cc	1090 calories/ gram	1803 calories/cc	64% of HMX

Sometime in the 1970s Soviet chemists developed a new explosive called ammonium dinitramide. This discovery remained secret until the molecule was rediscovered in the United States in 1989. On June 18, 1990, Jeffrey C. Bottaro and three other chemists at SRI International filed a patent on it. The molecule, nicknamed ADN, is a powerful oxi-

Ammonium dinitramide (ADN)

dizer and is being considered for new rocket propellants as a replacement for the troublesome ammonium perchlorates currently in use.

AMINES, AZIDES, AND NITRIDES

The amines are another class of explosives with a large number of compounds. I have already discussed some special cases of amines, such as the azides, and the tertiary amines nitrogen trichloride and nitrogen triiodide. Nitrogen tribromide is another tertiary amine. Selenium nitride SeN is a highly explosive red powder. A yellow form, Se_2N_2, is also explosive, detonating at 200° C. Also explosive are Se_4N_4 and Se_4N_2. None of them has found any commercial or military uses.

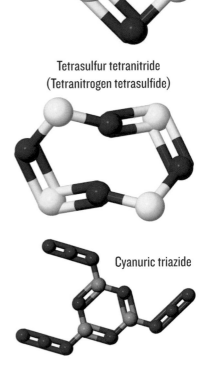

Disulfur dinitride

Tetrasulfur tetranitride
(Tetranitrogen tetrasulfide)

Cyanuric triazide

Chlorine azide

Sulfur also has explosive nitrides, notably disulfur dinitride, S_2N_2, and tetrasulfur tetranitride, S_4N_4. Both have interesting structural formulas.

Each of these compounds can have either the sulfur or the nitrogen first in the name, so disulfur dinitride is also dinitrogen disulfide.

Lead azide, silver azide, and sodium azide have been discussed earlier. They are all sensitive primary explosives used in percussion caps and airbags. Cyanuric azide is a heat-insensitive primary explosive with a respectable detonation velocity of 7,300 meters per second, similar to nitrocellulose or picric acid.

It is extremely shock sensitive, and the reaction products are graphite and nitrogen gas.

Chlorine azide is so unstable it can spontaneously detonate even when very cold. This makes it useless. But as a bookend to nitrogen trichloride, I include it here.

Other uselessly sensitive explosive azides are copper(II) azide, ethyl azide, the gas fluorine azide, silicon tetraazide, and tetraazidomethane.

Tetraazidomethane

Another sensitive azide is 1-(5-tetrazolyl)-3-guanyl tetrazene hydrate, known as "tetrazene explosive" to distinguish it from the simpler, nonexplosive molecule tetrazene. It is more sensitive than mercury fulminate and, like that compound, it is used as a primary explosive in detonators. It is not to be confused with tetracene.

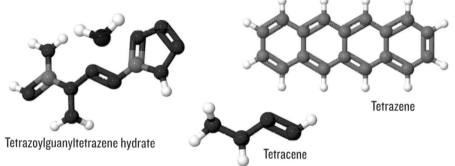

Tetrazoylguanyltetrazene hydrate

Tetracene

Tetrazene

In 1999 chemist Karl Otto Christe created the first new all-nitrogen molecule in over a century. The new molecule, N_5^+, is the third all-nitrogen species known, after N_2 (the nitrogen gas in the air) and azide, N_3^-, the part of azides that makes them go bang. N_5^+ is now known as pentazenium.

Pentazenium

Christe was working for the Air Force Research Laboratory at Edwards Air Force Base in California. The lab was working on alternatives to hydrazine, the highly toxic propellant used in many of the air force rocket engines. Not at all stable by itself, pentazenium combines with antimony hexafluoride to form $N_5^+SbF_6^-$, which is shock resistant and heat stable up to 60°C.

Finally, now that cubane has been synthesized, and octanitrocubane produced, the challenge for chemists is to come up with the next all-nitrogen species that might actually be stable, octaazacubane. It has not (as of this writing) been synthesized.

Octaazacubane

Similarly, an all-nitrogen version of buckminsterfullerene, N_{60}, would be a very high-density, possibly insensitive, high explosive that would produce only nitrogen gas as a reaction product. It is on a lot of wish lists.

PEROXIDES

N_{60}

The next group of explosives is the peroxides. Triacetone triperoxide (TATP), benzoyl peroxide, peroxymonosulfuric acid, and methyl ethyl ketone peroxide (MEKP) have been discussed earlier. Like them, the molecule hexamethylene triperoxide diamine (HMTD) is an unstable primary explosive that easily detonates with heat, impact, or friction. Because it is easy to make, it is used by suicide bombers and other terrorists (or would-be terrorists). It may be that once one has decided on suicide, the timing is not that important.

FURAZANS AND FUROXANS

Furazan

The molecule furazan (1,2,5-oxadiazol) has some nice properties as a starting point for building a new explosive molecule.

It is flat, which means it can pack more easily in a crystal, giving it a high density. It is stable. Ring molecules with alternating double and single bonds share the electrons between those bonds, so it acts as if each bond has one and a half electrons. This makes these "aromatic" rings more stable than other structures. The oxygen atom in the furazan is not bonded to carbon, but to two nitrogen atoms. This is a weaker bond, allowing the oxygen to free up in an explosion and combine with hydrogen or carbon, producing energy. The two nitrogen atoms pair up to form nitrogen gas.

Furazan was first synthesized in 1964. By 1968 Michael Coburn (the creator of the explosives NTO, PATO, PYX, and ATNI) had synthesized several furazan compounds, such as DAF, or diaminofurazan. Other chemists added to the furazan list, creating ANF, DAAF, and DAAzF in 1981 and DNFX in 1994.

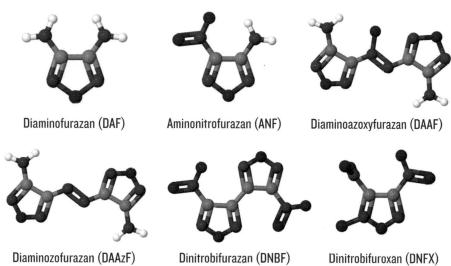

Diaminofurazan (DAF)   Aminonitrofurazan (ANF)   Diaminoazoxyfurazan (DAAF)

Diaminozofurazan (DAAzF)   Dinitrobifurazan (DNBF)   Dinitrobifuroxan (DNFX)

DAAF is used as a ballistic modifier in ammonium perchlorate composite propellants (solid rocket fuels). As DAAF decomposes, it releases ammonia, which slows the burning rate of the ammonium perchlorate fuels. DAAF is also being investigated as a less sensitive replacement for PBXN-7 (60% TATB, 35% RDX, 5% Viton) and PBXW-14 (50% HMX, 45% TATB, 5% Viton). Matching the shock sensitivity of these explosives was an important factor.

DAAF is also being looked at in castable replacements for Composition B (RDX and TNT) for environmental reasons. Its synthesis is environmentally friendly, and it is suitably insensitive, while acting in other ways identically to Composition B when it is combined with melt-castable low-melting-point explosives like ammonium nitrate and urea (in a combination with a lower melting point than either ingredient alone—called a *eutectic mixture*).

Furazans can be oxidized to create furoxans. DNFX is a furoxan. It is fully nitrated and highly oxidized. DNFX is a liquid that is stable at

−20°C but decomposes slowly at room temperature. It is very sensitive and must be handled carefully. However, as a starting point for other furoxan explosives, it has reactive nitro groups that can be substituted, making larger, more stable molecules.

One very energetic furoxan is dinitrodiazenofuroxan, or DNAF. With a detonation velocity of 10,000 meters per second, it may be the current record holder for an energetic molecule, on a compara-tive basis with CL-20 and octani-

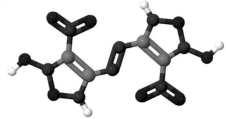

Dinitrodiazenofuroxan (DNAF)

trocubane. But it has a high impact sensitivity, and decomposes at 127°C, so as a practical explosive, it does not compare to those two.

As an example of a typical furazan, let's look at the molecule dinitrohydrazinofurazan.

It has two carbon atoms, two hydrogen atoms, five oxygen atoms, and six nitrogen atoms. Using the earlier rule of thumb for figuring out what the reaction products are, first remove all the nitrogen atoms and then pair them up. There are three pairs. Next, take pairs of hydrogen atoms and add an oxygen atom to each one to form water vapor. There are only two hydrogen atoms, so there is only one water molecule. Now take the carbon atoms and add an

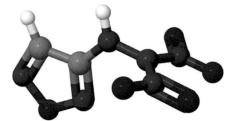

Dinitrohydrazinofurazan

oxygen atom to each one to make carbon monoxide. There are two oxy-gen atoms left over, which can be used to oxidize the carbon monoxide molecules to carbon dioxide.

One molecule of dinitrohydrazinofurazan will release three N_2 mol-ecules, one water molecule, and two carbon dioxide molecules. There is nothing left over. It has a perfect oxygen balance.

THERMOBARIC EXPLOSIVES

Up to this point, I have discussed explosives that contain their own oxidizer or have no oxidizer (they simply break up into simpler molecules). But in many military applications there is plenty of oxidizer around, in the form of the oxygen in the air. Not having to carry around all of the oxygen means the weapon can be smaller, and thus easier to carry around, perhaps even man-portable, or huge, consisting of a tanker-load of fuel.

The term *thermobaric weapon* has been applied to this type of explosive. It refers to the heat and pressure they produce, as opposed to the shock wave of a conventional explosive.

As with gunpowder, mixing the fuel and the oxidizer very closely is key to making an explosion. In the case of gunpowder, even that was not enough; the mixture had to be confined for a while until it could build up enough pressure to suddenly burst the container, in the case of a bomb, or move the projectile, in the case of a gun. In the open air, containment is obviously a problem. But in a cave or a bunker, a mix of fuel and air can have the same explosive effect seen in mine explosions, where methane or coal dust mixes with air in a confined space.

If the amount of fuel is large enough, and it is well mixed with just the right amount of air, the containment issue is less relevant. The target will be hit from all sides, and the inertia of the blast will produce its own confinement.

In a typical fuel-air bomb, the fuel (such as a hydrocarbon like gasoline) is spread into the air by one conventional explosive, and then ignited by a second explosive that is timed to allow the fuel and the air to properly mix. The fog that results from the first explosion can drift into foxholes and buildings, where it will be confined until the second explosion ignites it. Thus, these types of weapons are very effective in destroying bunkers, caves, and buildings, and any people in them.

If the fuel is preheated enough by the first explosion, usually by enclosing it in a strong casing to delay the actual spread until the fuel has been heated, it may be above its ignition temperature when it encounters the air, and no second ignition charge is needed.

One thermobaric weapon, the Russian Aviation Thermobaric Bomb of Increased Power (ATBIP), is thought to use powdered aluminum and ethylene oxide as the fuel. There are 7.1 tons of fuel in the bomb, which delivers the explosive equivalent of 44 tons of TNT when detonated. The blast radius is 1,000 feet. Due to secrecy and the propaganda value of a weapon such as this, the details are not easily verifiable.

The US BLU-118/B thermobaric weapon was designed to destroy tunnel complexes in Afghanistan. It could be detonated at the entrance to the tunnel, or set to detonate after skipping on the ground into the tunnel, or set to detonate after penetrating the tunnel from above. Insensitive explosives are needed for the last two methods, as the bomb takes a beating getting through the rock above the tunnel, or skipping on the ground after being dropped from a jet.

The insensitive explosive used in the BLU-118/B was PBXIH-135, a mix of HMX and fine aluminum powder in hydroxyl-terminated polybutadiene (a polyurethane rubber used to make skateboard wheels and Spandex). The fuel was a solid fuel, mostly aluminum powder. The BLU-118/B is a single-stage detonating device, and relies on the confinement inside the tunnel to increase the shock wave of the explosion. The explosion begins small but builds inside the tunnel, and lasts longer than a conventional explosive detonation would.

In tests in tunnels in Nevada, the bomb worked as designed. In actual use in Afghanistan, the first use missed the target, hitting a ridge in front of the tunnel instead. Since the explosive was smaller than a normal bomb (not counting the solid fuel), it did less damage in the open-air miss, which is considered an advantage of the weapon.

The BLU-73/B was a fuel-air explosive used in the Vietnam War by the United States, developed at the China Lake facility. It weighed 100 pounds and contained 75 pounds of ethylene oxide fuel. Its fuse was set for 30 feet above the ground, creating a cloud of fuel 60 feet in diameter. This was then ignited by an embedded detonator. The BLU-73/B was a sub-munition carried in the 550-pound cluster bomb called the CBU-72/B. The US Marine Corps dropped 254 of these from A-6-E attack jets. Designed to attack trenches and clear minefields, it is unclear if it was actually effective at clearing mines.

ECO-FRIENDLY EXPLOSIVES

Even when they aren't blowing up the countryside, explosives and rocket propellants can be harmful to the environment.

Ecological damage is expected in war, but as a side effect. Minimizing ecological damage is important for many reasons. First, after hostilities cease, people need to live, work, and farm in the areas previously engaged in conflict. There is also the effect that harmful or toxic compounds have on the troops using the munitions. Winds, water currents, and groundwater supply contamination can affect noncombatant nations, including allies and the nations using the munitions.

But even if there were no hostilities, the mere production of the munitions can have ecological consequences, as can the testing of them and their use in training and target practice. Moreover, when bombs and shells age or become obsolete, they must be disposed of in a way that does not pollute.

Explosives are also used in mining and oil and gas exploration, where their products are not only in the air the workers breathe but can pollute groundwater supplies.

The two main ingredients in primers for military, police, and civilian firearms are lead azide and lead styphnate. These were selected in the early 1900s to replace mercury fulminate and eliminate the toxic effects of mercury. Unfortunately, lead is also a toxic heavy metal, and industries are just now getting around to eliminating *it*, in turn.

Toxic lead salts at firing ranges can reach dangerous levels. One report in 1991 found that lead in the cells of people who had just cleaned an FBI firing range was over ten times above government health limits. Over a thousand pounds of lead primary explosives are used by the US Army alone each year (710 pounds of that is lead), just from the primers in small caliber ammunition. Ninety-five percent of that is used in training, meaning that the lead pollution is here at home and primarily affects our service personnel and police officers. (Mercury fulminate is currently illegal in most of the world for use in primers and blasting caps.)

Replacing primer ingredients with safer compounds is ongoing. While it might seem to be a trivial change, one US Department of Defense study found that primers made with DDNP (diazodinitrophenol, discovered in 1858 by Peter Griess, the same person who later discovered lead styphnate) had perceptible delays (seventy-four milliseconds) in igniting the secondary explosive in the ammunition tested, while lead-based primers in the same systems did not. These delays contributed to losses in accuracy in the weapons and at least one misfire. The DDNP primers had more variation in peak blast pressure and did not adequately ignite the secondary charge.

As these tests were on one batch of Russian-made primers, it does not follow that DDNP is a failed explosive when used as a primer. Manufacturing variances and light loading may explain the performance. Nevertheless, simply replacing primer compositions is not a simple matter and requires quite a bit of testing and manufacturing expertise with the new formulas. There is some evidence that the performance deficits from DDNP are not due to the explosion of the primer, but due to the effect that lead styphnate has on reducing bullet friction in the gun barrel. Without the extra lubrication the lead-based primer supplies, the performance in the actual weapon is erratic. Testing of the primers in the lab would not discover this.

Lead is not the only toxic substance in primers. Some use barium nitrate oxidizers, and some contain antimony. While the emission of heavy metal salts as fine particulates in the air during use is one issue, phasing out toxic materials throughout the supply chain is also desirable. Factories that produce the ammunition do not wish to pollute or expose their workers to toxic materials.

At the US Naval Air Warfare Center in China Lake, California, work is under way on electrically ignited primers. A mix of aluminum nanoparticles and carbon black, with molybdenum trioxide as an oxidizer, is electrically conductive due to the first two ingredients. The goal is ignition within four milliseconds of the current being switched on. Called *nano-thermites*, these combinations of aluminum powder and oxides of less reactive metals such as iron, copper, or molybdenum are blended together so that the aluminum and the metal oxide are joined together in particles smaller than 100 nanometers. This leads to very fast reaction times.

The US Naval Surface Warfare Center in Indian Head, Maryland, is also working with nano-thermites. Called metastable intermolecular composites (MICs), or metastable nanoenergetic composites (MNCs), the names refer to the fact that they are stable as aluminum and metal oxide but can flip to aluminum oxide and elemental metal (releasing a *lot* of heat—thermite is used to melt steel). In their testing, the researchers are using bismuth oxide, Bi_2O_3, as the metal oxide. It is being considered for all small and medium caliber ammunition up to calibers of 40 millimeters.

In addition to bismuth oxide, the Indian Head researchers are also experimenting with molybdenum oxide, tungsten oxide, and copper oxide. One problem with the materials is that they are sensitive to friction and electrostatic discharge (static electric sparks). For safety, the mixture is prepared wet and dried only once it has been pressed into the primer cap. The mixing of the aluminum and the metal oxides is done using ultrasonic mixing equipment, and the aluminum nanoparticles are coated with a protective layer so they do not spontaneously combust in air and so they can be mixed with water.

At the High Explosives Science and Technology Group in Los Alamos, New Mexico, chemist My Hang Huynh (who won a MacArthur Foundation "Genius" grant for her work in 2007) has been working with modifications of nitrotetrazole and its salts as a primer candidate.

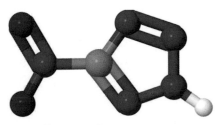

Nitrotetrazole

The resulting primer is insensitive when wet, an advantage from a safety point of view. Copper salts of the acid are environmentally benign. Potassium, rubidium, and cesium salts also work but do not meet the criterion of heat stability, as they decompose before the 200°C limit.

Primers are not the only problem, of course. The bullets are still lead. While shotgun loads have long been available with lead-free shot, made from bismuth (a dense, nontoxic element)

Copper(I) 5-nitrotetrazolate

or steel, replacing lead in bullets results in added costs or decreased performance. The current lead-free bullets being tested are copper-jacketed bullets filled with a mixture of tungsten and tin. Tungsten has a very high density, and the bullets can be made to have the same density as the lead bullets they replace (at a cost, however, since lead at 71 cents a pound is much cheaper than tungsten at $20 a pound). Costs are offset by the fact that metal prices are not always the gating factor in the price of ammunition, even in military quantities. Another cost savings is that the bullets can be easily recycled just by heating them. The tungsten powder is not wetted by the tin binder, and the two are easily separated.

TNT in the soil of battlefields and testing ranges is classified as a potential carcinogen. While TNT has long been disfavored in respect to nitramine-based explosives in militaries, it is still widely used in mining, for cost reasons. Mixing RDX or HMX with newer castable explosives like TNAZ results in the ability to tailor non-TNT explosives to match or surpass the performance of TNT. Again, nonpolluting solutions come at a cost. TNT is used because it is cheap.

Rocket propellants that use ammonium perchlorate (such as the space shuttle's solid rocket boosters) produce hydrogen chloride gas when burned, which becomes hydrochloric acid when it reaches water in the air or in lungs. The perchlorate itself causes thyroid problems and has been detected in groundwater near plants that produce it. Replacing perchlorates with environmentally friendlier propellants and oxidizers is a task that has been taking years, due to the testing and formula refinement required. This will be covered in more detail in the next chapter, as will the effects of chlorine-based propellants on the ozone layer.

HIGH-ENERGY ROCKET FUELS

For the ancient Chinese, a high-energy rocket propellant was black powder. As rocket technology improved, better propellants were found, often by borrowing from gun propellants. Gun propellants, such as guncotton and nitroglycerin, burn rather than detonate when not highly confined, and this is a good thing, since the last thing you want your rocket to do is blow up (and I guess, technically, that *would* be the last thing it did).

Rockets carry their fuel and oxidizer with them. This means that they must not only accelerate the payload but the fuel as well. This puts a premium on fuels with high energy per unit weight.

As with gun propellants, rocket fuels can be either a mix of oxidizing agent and fuel, as in black powder, or they can combine the fuel and the oxidizer on a single molecule, as with guncotton or nitroglycerine. The Apollo lunar missions used liquid hydrogen and liquid oxygen. The reactant was water vapor. The light weight of the water molecule (18 atomic mass units) is a benefit, because one of the constraints on rocket engine design is the speed of the reaction mass being thrown behind. The faster the exhaust gases, the better the performance. For a given amount of energy, lighter molecules move faster than heavier ones.

There are several problems with cryogenic liquid fueled rockets. The fuels have to be stored separately from the rocket. This is not a big problem for the Apollo program, but for systems that launch more frequently or need to be stored in an aircraft, storing the fuels in the rocket is desirable. The fuels need to be kept very cold. This means that cryogenic Dewars (and their associated weight) must be on board the rocket. The Dewars also need to be strong enough to withstand the forces the rocket undergoes during launch but also during shipping and outfitting. The fuels are low density. This means that large tanks are needed to store them. Large tanks add to the weight. Every bit of added weight means more fuel is needed, which adds more weight.

Solid propellants solve several of these problems but add their own costs. Solid propellants can be stored in the rocket. They don't need cooling. They are high density, so the container is smaller. On the cost side, the reactant gases are generally heavier molecules, such as carbon dioxide (44 atomic mass units), nitrogen gas (28 atomic mass units), and in some cases metal oxides or chlorides, which are even more massive. As a result, the exhaust gases are heavier and are harder to get moving at the same speed as water molecules.

The US Titan missile project was started in 1959. The Titan I used kerosene as the liquid fuel and liquid oxygen as the oxidizer. Each molecule of kerosene produced twelve molecules of carbon dioxide and thirteen molecules of water vapor.

By changing the fuel to dimethylhydrazine and the oxidizer to nitrogen tetroxide, the need for cryogenic storage was eliminated. The new Titan II rocket could have the fuel stored in it, allowing the missile to be fired in a single minute (hence the name Minuteman Missile). The simplified maintenance and storage meant the missile could be kept in a missile silo.

Those fuels were difficult to handle without accidents and led to a program to develop solid propellants. The Titan III missile, with solid rockets, launched in 1964. The fuel was aluminum powder, the oxidizer was ammonium perchlorate, and the two were mixed with polybutadiene acrylic acid as a binder. Aluminum powder is used because it generates a lot of energy when it burns (per unit weight), and it is denser than hydrocarbon fuels, so it takes up less space, meaning the container weighs less. It is also difficult to accidentally ignite, and it helps to stabilize the burn rate. Some mixes add iron oxide to increase the burn temperature through the thermite reaction.

The Minuteman is a three-stage rocket. The first stage uses the Titan III mix as described above. The second stage uses a similar mix with a different binder. The third stage adds HMX, nitrocellulose, and nitroglycerin to the mix. The higher energy molecules in the third stage allow the exhaust gases to reach higher speeds, improving the performance of the rocket. At low speeds (such as at liftoff) the exhaust gases are moving much faster than the rocket is, and so their speed is of less importance. As the rocket gains speed, high-speed exhaust gases become more important. The third stage is moving so fast, the cheap propellant is not cost effective, and the higher energy mix is needed.

The Trident missile system used a fuel called XLDB-70, which stands for "cross-linked double-base, 70% solids." The solids were HMX, aluminum, and ammonium perchlorate. The binder was polyglycol adipate, nitrocellulose, nitroglycerin, and hexadiisocryanate, making up 30% of the fuel. Cross-linking the binder means that extra chemical bonds are made in the binder, so that it is more stable, and stiffer. The polyglycol adipate was replaced by polyethylene glycol in the later Trident systems, and the new fuel was named NEPE-75. The 75 referred to the fact that with the improved binder it could hold 75% solids.

The use of ammonium perchlorate was found to cause environmental pollution. It produces hydrogen chloride gas when it burns (which turns into hydrochloric acid when it encounters water). Exposure to perchlorates over time produces thyroid problems. This is not just a problem when the missile launches, but it is a considerable problem for the companies making the fuel.

Besides the environmental impact of the perchlorates, the use of HMX in the Minuteman third stage and the Trident is a cause of concern because it is susceptible to detonation by shock or friction. Again, not only is this a problem for the missile, it is a problem for the companies working with the fuels.

Modern impact, friction, and heat insensitive high-energy molecules can be added to solid rocket fuels to increase safety and performance. Furazan-based molecules such as diaminoazoxyfurazan (called either DAAF or DAAOF depending on the researcher) have been added to mixes, with interesting results. The molecule adds high energy content, but also modifies the reaction rate of perchlorate mixes, helping to control the rate of burning. In mixes without perchlorate, its susceptibility to shock (it is being considered as a booster explosive for less sensitive explosives) may be moderated by the plastic or rubber binder.

One way to increase rocket performance is to increase the pressure. This makes the exhaust gases move faster. With modern lightweight but incredibly strong composite materials, increasing the pressure is definitely an option. The problem has been that as the pressure rises, the solid propellant burns faster. This in turn increases the pressure, leading to a runaway situation that ends in an explosive catastrophe.

DAAF can be added to perchlorate mixtures to modify the burn rate. In a similar fashion, perchlorate itself can be used (as much as 10% of a mix) to moderate nitramine propellants, such as a mixture of RDX, nitroglycerin, trimethylolethane trinitrate, and polyurethane. In these mixes, the pressure exponent (a measure of how quickly increased pressure causes the fuel to burn faster) goes from numbers greater than 0.9 to numbers less than 0.65, a very significant drop.

The nitramine-based propellants replace nitrocellulose and nitroglycerin double-base propellants, which has the benefit of reducing the smoke from the missile, making it harder to track. This is known as a "reduced signature" propellant for this reason. Low signature missiles fired from aircraft carriers or field emplacements do not give away the position of the attacker. Ammonium perchlorate propellants produce even more smoke in the exhaust than double-base propellants, giving another reason to phase them out in favor of higher energy molecules.

Propellants using CL-20 have even higher performance than RDX-based propellants, matching that of liquid-fueled rockets. As newer high-energy molecules are developed, this trend is expected to continue.

Replacing the inert binders with high-energy polymers such as GAP (glycidyl azide polymer, discussed earlier) increases the performance still further. By adding 1 to 5% heavy metal catalysts (lead, tin, or copper compounds), the pressure exponent can be dropped by large amounts. Further drops can be had by adding carbon in the form of carbon black, or graphite, in amounts less than 1% by weight. The pressure exponent in mixes like this is generally less than 0.6, and the chemical stability improves as well.

Combinations of the insensitive high explosive FOX-7 with GAP are being used in shoulder-launched rockets, replacing RDX and HMX, resulting in munitions that are safer to transport and handle, and less sensitive to detonation in a fire, or when hit by bullets, fragments, or nearby exploding ordnance concussions. However, FOX-7 is still (at the time of this writing) expensive and in limited supply.

Rocket fuels are rated by a number called the *specific impulse*. Specific impulse is a way of combining the performance-related metrics discussed earlier (heat of the reaction, temperature of the reaction products, and the molecular weight of the reaction products) into a single number that can be used to easily compare two fuels.

The units (Newton seconds per kilogram) are unimportant for this discussion, as the specific impulse values can be treated as simple numbers for comparisons. Fuels that burn hotter will have larger values for specific impulse. Fuels that contain more light elements such as hydrogen will have higher value than fuels that have metal oxides or chlorides in the reaction products, or fuels that produce a higher percentage of carbon dioxide.

Thrust, the force that moves a rocket, is a function of the specific impulse times the density of the propellant, times the rate of burning, times the surface area being burned. Increasing any of these factors increases thrust. Increasing the surface area is generally not feasible in a rocket due to the limited volume available (increasing surface area in a constant volume means using less propellant). Where high burn rates are more important than long burn times, such as interceptor missiles, increasing surface area by making star-shaped cavities in the propellant is a tradeoff that has benefits.

Burn rate modifiers can speed up the burn rate or slow it down, as seen with DAAF and perchlorates. Adding iron oxides to fuels that contain aluminum raises the temperature of the reaction through the thermite process. Adding heat conductive metal whiskers can also speed up the burn rate.

Thus, two other important metrics for judging rocket fuels are the burn rate and the associated pressure exponent. The rate at which a propellant burns is proportional to the pressure raised to the pressure exponent. Exponents less than one mean that as the pressure increases, the burn rate increases, but at progressively slower rates.

The lowest pressure exponent fuels are the same as those used in guns— double-base propellants containing a mix of nitroglycerin plasticized with nitrocellulose. They have pressure exponents in the range of 0 to 0.3, meaning that as the pressure increases, the burn rate increases proportionally, with no tendency toward runaway explosions. Unfortunately, they also have the lowest specific impulse (2,100 to 2,300), due to the relatively cool temperatures and the carbon dioxide they produce. Their burn rates are in the 10 to 25 millimeter per second range.

Next are the ammonium perchlorate–based fuels. Those containing nitramines and aluminum have specific impulses in the 2,500 to 2,600 range, but they produce the most smoke. The reduced signature versions without the aluminum have specific impulse values of 2,400 to 2,500. The pressure exponents are in the range of 0.3 to 0.5 (an exponent of 0.5 means that the burn rate goes up as the square root of the pressure). The burn rates are in the 6 to 40 millimeter per second range.

Ammonium nitrate–based propellants have slightly lower specific impulse values (2,200 to 2,350) and pressure exponents of 0.4 to 0.6. They are low signature and generally consist of ammonium nitrate, nitramines, and GAP; the burn rates are 5 to 10 millimeters per second.

Also in the same exponent range (0.4 to 0.6) are mainly nitramine-based propellants with GAP and some ammonium perchlorate (less than 10%) to bring the pressure exponent down. They have specific impulse values in the 2,300 to 2,450 range.

The latest solid rocket propellants contain ammonium dinitramide (ADN) as the oxidizer instead of ammonium perchlorate. ADN is a high-energy molecule by itself with a respectable detonation velocity of 8,074 meters per second (close to TATB). When burned, it produces more gas than any of the other nitramides, even more than RDX. As an oxidizer, it has an oxygen balance of 25.8, which puts it in between ammonium nitrate (20.0) and ammonium perchlorate (34.0). Because it burns hotter and has lighter reaction products, it can perform better than ammonium perchlorate. Its other benefits as a replacement for ammonium perchlorate are that it is low signature (less smoke) and it does not produce any hydrogen chloride gas (just water, CO_2, and nitrogen gas).

Mixtures of ADN, RDX, GAP, and aluminum reach specific impulse values higher than any of the previous solid propellants, in the 2,500 to 2,700 range.

Another high-energy oxidizer used for replacing ammonium perchlorate is hydrazinium nitroformate, or HNF.

Mixtures of HNF with aluminum and GAP, polyGLYN, or polyNIMMO are showing promise as high-energy propellants with reduced signature, high specific impulse, and no chlorine in the exhaust. Chlorine, in addition to pro-

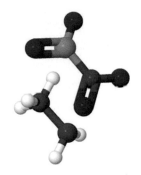

Hydrazinium nitroformate (HNF)

ducing hydrochloric acid when burned with hydrogen-containing fuels, is a contributor to ozone depletion in the upper atmosphere, which is especially important when dealing with upper stage booster fuels. The closer to the ozone layer the chlorine is released, the more pronounced the effects.

A competitor to ADN/aluminum is CL-20 mixed with energetic binders such as GAP. CL-20 is currently available in kilogram quantities, but production is expected to increase as more uses for the very energetic molecule are found. Made in batches of 50 to 100 kilograms, it is still expensive and less easily available than the ADN propellants. It is also more sensitive to heat, shock, and friction.

TNAZ is also being tested in rocket propellants. Its advantages are an energy content close to RDX, a melting point that makes it castable like TNT, and chemical compatibility with metals like aluminum and steel, as well as the common binder polymers and plasticizers.

I have been assuming in the foregoing discussion that a pressure exponent of 0.7 or above is to be avoided. This is to reduce the possibility of a runaway reaction that eventually exceeds the burst strength of the combustion chamber.

However, there is a class of rocket engine where a higher pressure exponent is desired or even required. This is the *thrust magnitude controlled* class (TMC), an engine where gases or fluids are injected into a solid rocket combustion chamber to change the thrust. A propellant with a high pressure exponent allows a small change in pressure to have a larger change in the burn rate. The thrust can thus be controlled with smaller changes to the flow rate of the gas or fluid being injected.

TMC rocket engines are being used in upper stage rockets for satellite launch and spacecraft propulsion systems.

A typical TMC engine might inject the water vapor and oxygen provided by decomposing hydrogen peroxide. The decomposition can be easily controlled, and the resulting gases have enough pressure to enter the combustion chamber. There they expand and possibly react with a low oxygen balanced propellant, increasing the pressure. The high pressure exponent causes the burn rate to quickly rise.

Since the pressure is under the control of the injection system, runaway is not an issue. The pressure can be kept below any critical limits by reducing the flow rate.

As with the move toward less sensitive explosives, safer rocket propellants are also a high priority in military rockets. A fire on an aircraft carrier deck, or enemy bullets and shell fragments, should ideally not set off rocket fuels any more than they should explode bombs.

So-called LOVA propellants (for low vulnerability ammunition) should not unintentionally burn or explode, and should not release toxic combustion products. Two of these propellants are XM39 and M43. They both consist of 76% RDX, 4% nitrocellulose, and 0.4% ethyl centralite. The latter compound is a plasticizer used in many smokeless powder gun propellants.

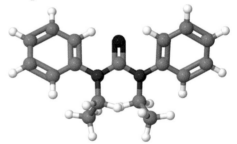

Ethyl centralite

The remaining ingredients are 7.6% acetyl triethyl citrate, known as ATEC (a plasticizer used in cellulose derivatives and PVC plastics) in XM39, and an (unspecified) energetic plasticizer in M43 (energetic meaning it has azide or nitro groups added to make it higher energy).

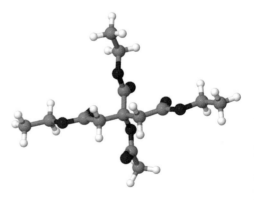

Acetyl triethyl citrate (ATEC)

A BRIEF HISTORY OF EXPLOSIVES
IN THE HOME

From the gasoline explosions that move those old-fashioned nonelectric cars to the match that lights the fire in the fireplace, explosives have been a part of daily life for most of the world's population for some time.

The mixture of chemicals in a kitchen match is in many ways quite similar to black powder. You can smell the sulfur, and from the rapid flare of the first strike, you can tell there is a powerful oxidizer in the mix. To get it started, there is a little extra something: the element phosphorus in some form.

That form makes the difference between safety matches and "strike-anywhere" matches. In a safety match, the phosphorus is on the outside of the box or book of matches, and is in its stable red form mixed with glue and powdered glass. The glass is there to add friction and sand off tiny particles of the match head when it is struck.

The red form of phosphorus is the amorphous form, where the bonds between the atoms are relaxed and unstrained. The more reactive white form of phosphorus is a tetrahedron of four phosphorus atoms where the bonds are strained and store energy that can be released easily.

Tetraphosphorus P$_4$

These strained bonds have appeared in several explosive compounds discussed earlier, in powerful explosives such as CL-20, octanitrocubane, and TNAZ. White phosphorus will burn in air at room temperature, producing the glow that gives the name to "phosphor" and "phosphorescence," even though today these effects are achieved using much safer molecules.

When the head of a safety match is rubbed against the striker, a tiny bit of the red phosphorus is heated by the friction to a temperature at which it changes form and becomes white phosphorus. This hot white phosphorus mixes with the air but also with the many tiny particles of sulfur and oxidizer from the match head. The match head catches fire and flares strongly until all the oxidizer is used up.

The oxidizer is potassium chlorate. In 1787 the French chemist Claude-Louis Berthollet bubbled chlorine gas through a hot solution of potassium hydroxide and was the first to create this compound. It is still sometimes referred to as Berthollet's Salt. He found that it exploded violently when mixed with carbon, which made him think it would be a good idea to make gunpowder with it instead of saltpeter. The result is so unstable that the first public demonstration of the new chlorate gunpowder killed five people.

In the match head, this instability helps to start the fire. When mixed with ingredients that are very easy to ignite, like sulfur or antimony trisulfide

Antimony trisulfide

(another ingredient sometimes used in matches, despite its toxicity), it is so easy to ignite that one can light a match with a flick of a finger.

Adding phosphorus to potassium chlorate produces a primary explosive so easy to detonate it is used to make paper percussion caps for toy cap guns.

In the safety match, the phosphorus is kept separate from the chlorate until the user brings them together by striking the match. In strike-anywhere matches, the phosphorus is in the form of phosphorus sesquisulfide.

Phosphorus sesquisulfide was first prepared by the French chemist Georges Lemoine in 1864. In 1898 Henri Sevene

Phosphorus sesquisulfide

and Emile David Cahen, working for the French government's match monopoly, first used it in matches, to replace the white phosphorus that was previously used.

Matches had become such a staple of life that the production of them was not only a monopoly of the government but was a significant health problem. The white phosphorus was causing a debilitating disease called "phossy jaw." Once a workable substitute had been found, governments around the world started banning (or in the United States, punitively taxing) matches made from white phosphorus.

The tips of strike-anywhere matches are actually an explosive. You can detonate them by hitting them with a hammer. They are mostly phosphorus sesquisulfide and potassium chlorate with a little binder and some glass powder, which absorbs heat and stays hot, thus helping to keep the reaction going.

The rest of the head is the same but with some of the phosphorus sesquisulfide replaced with sulfur. There is some wood rosin and a bit of paraffin to provide fuel to keep the match burning. That fuel is needed because the wooden sticks have been coated with a fire retardant (ammonium phosphate) so the match goes out before it reaches a finger.

Safety matches have antimony trisulfide and sulfur mixed with the potassium chlorate, glue, and powdered glass. The antimony trisulfide is easier to ignite than sulfur, and safety matches need some help, since the phosphorus is on the box, not in the match head.

The first matches, like the first black powder, were made in China. Pinewood sticks coated in sulfur were used to restart cooking fires. They had to be lit by pressing them into a hot coal from the damped fire of the day before, but they saved time that would otherwise be spent blowing on the hot coal after adding some other flammable material. The sulfur was flammable enough to work without the extra wind.

In 1669 the alchemist Hennig Brand discovered phosphorus by boiling down 1,500 gallons of urine in an effort to make silver. He had gone to the German army to get permission to collect urine from all the soldiers at the camp. The phosphates in the urine gave their oxygen to the carbon from organic matter in the liquid, and elemental phosphorus was left over. Brandt kept the secret of his preparation in an effort to sell it. In

1677 Johann Crafft, on a visit to the polymath Gottfried Wilhelm Leibniz (known for inventing calculus), told the court of the Duke of Hanover about Brand's discovery of a cold fire. Leibniz writes that it can be rubbed on the face and clothes to create light (and "very pretty effects"), and that it does not ruin the clothes. (Don't try this at home—phosphorus burns are awful and take a long time to heal.)

Leibniz contacted Brand and told him the duke would like to hire him on a monthly basis to tell the court about his discoveries. Johann Kunckel, the alchemist I earlier discussed as the one to first discover mercury fulminate, had also heard of Brand through Crafft. Like Crafft, he also offered Brand money for his invention, but he made the mistake of mentioning his plans to Crafft. Crafft and Leibniz immediately offered Brand a substantial amount of money not to tell Kunckel the secret. Brand told Kunckel he was unable to reproduce his results. Much later, claiming to be fed up with Crafft, Brand came back to Kunckel offering to sell the secret.

By then Crafft had traveled to England showing off the new cold fire. He showed it to the English chemist Robert Boyle but would not tell him the secret of how to make it, instead just hinting that it came from "somewhat that belonged to the body of man." Whether it was that clue or the visit his assistant Ambrose Godfrey-Hanckwitz made to Brand in Hanover, Boyle was able to produce his own phosphorus. The key was the very high temperatures Brand had used. Boyle and Godfrey-Hanckwitz improved on the method, adding sand to take up the sodium from the phosphate, and Godfrey-Hanckwitz went on to produce it on an industrial scale.

Phosphorus was still being made from human urine until, in 1769, Swedish chemists Johan Gottlieb Gahn and Carl Wilhelm Scheele (having read Kunckel's papers) showed that bones are made from calcium phosphate, and extracted phosphorus from the ashes of bones. This discovery eventually made Sweden the world's leader in the manufacture of matches. Scheele went on to discover oxygen in 1771, and the elements barium, manganese, molybdenum, tungsten, and chlorine, before dying at age forty-three from heavy metal poisoning. He was known to sniff and taste any new substances he had discovered.

In the 1840s phosphates from bat guano deposits were used to make phosphorus, and by 1850 phosphate rocks were used, replacing the bone ash. The rocks contain calcium phosphate and produced phosphorus when heated with coke and sand in an electric arc furnace.

The first self-igniting match didn't bother with phosphorus, however. In 1805 chemist Jean Chancel invented it. A mix of potassium chlorate, sulfur, sugar, and rubber made up the match head. These were dipped into a little bottle made of asbestos that contained sulfuric acid. The reaction started the fire. A mix of potassium chlorate and powdered sugar ignited by sulfuric acid is still a common demonstration in chemistry classes.

The first friction match was invented by Frenchman François Derosne in 1816. His sulfur-tipped sticks were rubbed against the inside of a tube coated with phosphorus. A slightly more convenient and commercially successful friction match was invented in 1826 by English chemist John Walker. He again did not bother with phosphorus but used potassium chlorate, antimony trisulfide, sulfur, and glue. Without the phosphorus, he had to fold over a piece of sandpaper and pull the match through it. As anyone who has failed to strike a safety match the first time would expect, the results were not always reliable.

It was the French chemist Charles Sauria who thought of replacing the antimony trisulfide with white phosphorus. With less sulfur odor, these matches were more popular than Walker's were.

In 1850 Austrian chemist Anton Schrötter von Kristelli found that heating white phosphorus to 250°C (without oxygen) turned the white form into the amorphous red form, which did not have the toxicity of tetraphosphorus and did not fume and glow in air. However, it was more expensive than the white form, which was used in matches until Sevene and Cahen made their phosphorus sesquisulfide version.

The US patent for their match, filed on July 19, 1898, mentions older matches using red (amorphous) phosphorus:

To all whom it may concern:

Be it known that we, HENRI SEVENE and EMILE DAVID CAHEN, citizens of the Republic of France, residing at Paris, France, have invented Inflammable Paste for the Manufacture of Matches, of which the following is a specification.

Experience shows that matches that are sensitive at all their surfaces and are capable of being easily transported can be manufactured by employing mixed pastes containing a mixture of white phosphorus and an oxidizing body such as chlorate of potash, bichromate of potash, and the oxids of lead or of manganese; but these pastes present from the point of view of hygiene well-known inconveniences.

We have tried to substitute for white phosphorus in the mixed pastes a body which, while being harmless to the health of the workpeople, might enjoy its essential properties of possessing a definite chemical composition and being easily inflammable. The sesquisulfid of phosphorus fulfills these prime conditions. Moreover, it offers a sufficient resistance to moisture and to atmospheric agents. It can be manufactured industrially and obtained in the state of purity by distillation. The matches manufactured with mixed pastes containing this sesquisulfid, oxidizing bodies, inert matters, and glues are satisfactory and can furnish all the degrees of sensitiveness desired by slight variations in the relative proportions in which the materials are used. We will give by way of example the following composition: sesquisulfid of phosphorus, ninety grams; chlorate of potash, two hundred grams; peroxid of iron, one hundred and ten grams; zinc white, seventy grams; powdered glass, one hundred and forty grams; glue, one hundred grams; water, two hundred and ninety grams.

We are aware that efforts have often been made to employ in the preparation of mixed pastes for matches a mixture of amorphous phosphorus and sulfur either in powder or in the state of fushion but these mixtures do not answer the purpose sought. They differ essentially from the sesquisulfid of phosphorus that we employ in that this last body is a perfectly definite composition that is very stable, resists moisture, as well as the atmospheric agents, and can easily be utilized and manipulated industrially.

The reference to the "health of the workpeople" is in recognition of the strikes at English match factories in 1888 over the health concerns of working with white phosphorus.

Various countries banned white phosphorus matches in the years that followed, and on September 26, 1906, the Berne Convention banned them in all countries by international treaty. Manufacture was banned as well as imports and sales.

Gustaf Erik Pasch, a professor of chemistry in Stockholm, was granted a patent on the safety match in 1844. He had moved the phosphorus to a striking surface, away from the head of the match. When red phosphorus became available in the mid-1950s, the company making safety matches switched to it. This is the type of match we use today.

Moving on to more explosive devices, the toy cap gun was introduced in the United States by firearms companies that no longer had a huge market after the end of the Civil War. After World War II, with the advent of television westerns, the products became very successful.

The paper caps for the guns held a tiny drop of red phosphorus and potassium chlorate between a paper bottom sheet and a tissue paper top sheet. The mixture was less dangerous to manufacture than you might expect, because water was used to make a paste of the two ingredients, which were only allowed to dry later.

Later versions used small plastic cups in a plastic wheel. Some used a form of black powder with powdered zinc added to make the powder more impact sensitive.

Another explosive toy is the party popper. It also uses red phosphorus and potassium chlorate, ignited by a pull-string, and shoots tiny rolls of fireproof paper that unwind to festoon merry-makers at New Year's Eve parties.

The Big Bang Cannon uses calcium carbide powder to create acetylene gas when mixed with water in the cannon. A spark from a flint then ignites the gas, producing a very loud report and a flash of sooty orange flame.

A little toy that goes by various names, such as Snaps, Silver Torpedoes, Pop-Its, Throwdowns, snappers, and Bang Snaps, are little tissue-paper bags of sand, with a tiny amount (less than a tenth of a milligram) of silver fulminate coating the sand grains. They explode when thrown on the ground, are stepped on, or just rubbed between finger and thumb. The sand prevents the explosion from causing much damage by absorbing most of the impact.

Cracker Balls are the larger cousins of Snaps. Made with a fulminating gunpowder mix made from potassium chlorate, antimony trisulfide, carbon black, aluminum powder, and sand, they are manufactured in wet form for safety and dried slowly after being coated in a form of papier-mâché. When thrown on the ground or against a wall, they make a much louder explosion than Snaps do.

Estes rocket engines are cardboard tubes filled with a compressed black powder, with a clay nozzle pressed onto one end. They are ignited by a hot wire sometimes coated with a pyrotechnic mix of nitrocellulose and black powder.

Jetex was a small rocket engine for model airplanes. Developed in 1947, it was popular until the mid-1970s. They were aluminum jars with a small hole for a nozzle. The user filled them with a slug of guanidinium nitrate, an explosive that burns cool and produces a large amount of gas. As a jet engine for model airplanes, it featured an acceleration low enough not to rip the wings off the model and ran cool enough not to burn the balsa wood.

INDEX